Gardeners' World

ALMANAC

ALMANAC

A MONTH-BY-MONTH GUIDE
TO YOUR GARDENING YEAR

FOREWORD BY
MONTY DON

BBC Books, an imprint of Ebury Publishing
20 Vauxhall Bridge Road
London SW1V 2SA

BBC Books is part of the Penguin Random House group of companies
whose addresses can be found at global.penguinrandomhouse.com

Penguin
Random House
UK

This book is published to accompany the television series entitled,
Gardeners' World, first broadcast on BBC2 in 1968

First published by BBC Books in 2021

www.penguin.co.uk

A CIP catalogue record for this book is available from the British Library

ISBN 9781785947520

Publisher: Albert DePetrillo
Project editor: Joanna Stenlake
Design: seagulls.net

BBC Gardeners' World Magazine: Tamsin Hope-Thomson, Kevin Smith

Printed and bound in Great Britain by Clays Ltd, Elcograf S.p.A.

The authorised representative in the EEA is Penguin Random House Ireland,
Morrison Chambers, 32 Nassau Street, Dublin DO2 YH68

CONTENTS

Foreword vii

January 1

February 25

March 51

April 79

May 107

June 141

July 169

August 195

September 219

October 243

November 265

December 285

Acknowlegdments 307

Index 308

FOREWORD

Over the years I have kept a garden diary. Often this is simply a record of what jobs have been done, seeds sown, plants planted or harvests gathered. In themselves these diaries make for pretty prosaic reading. But over the years a pattern emerges of what happens when, of what has worked well and – just as importantly, what has not worked. Doing the right thing at the right time is not just good for the garden but helpful for the gardener, to ensure that time and energy are spent most effectively.

To every thing there is a season but if this becomes a tyranny then it is going wrong. It can sometimes seem as if there is a precise date for each particular garden task – which if missed, spells horticultural disaster and ignominy. The right thing at the right time, using the right technique, right tool, right place, right plant (preferably knowing its right Latin name) are all hurdles that the aspiring gardener can feel blocking the road to straightforward gardening pleasure.

Fortunately it really is not like that. The more you garden, the more you realise that almost everything is flexible and nuanced and the best gardens are a fluid interaction between the gardener and the garden – even if this might not be right for anyone else. There is no one absolute way to do anything but you can be confidant that if it works for you then you are doing it right.

However, I know that this kind of blithe assurance can be inhibiting and even annoying. The truth is that only confidence in a certain mastery of a subject allows the freedom to break horticultural conventions and rules. All of us need much more specific guidance from time to time and especially for those gardeners starting out, a clear guide through the year of what to do and when to do it is invaluable.

The right thing done at the wrong time can sometimes be as bad as the wrong thing done at the right time. All gardening involves planning across weeks, months and seasons. The purple sprouting broccoli harvested in April will have been sown a year earlier and planted out the previous June. Roses flowering in July are best planted in midwinter and wildflower seed that lights up spring should be sown in early autumn. It is not just a matter of planning your own gardening diary but also of knowing how certain plants behave across the year. Late-flowering clematis like *Clematis viticella* are best pruned very hard in early spring as they flower only on new growth, but treat the early-flowering ones, like *Clematis montana* or *C. alpina*, the same and you will have no flowers at all as they flower on growth made the previous summer. Spring-flowering bulbs need planting the previous autumn and biennials such as foxgloves or honesty must be sown in early summer in order to flower the following year while annuals do their geminating, growing, flowering and setting seed all within the span of one brief season of spring to late summer. Managing all this information and timing it right can feel overwhelming, with the potential for disaster taking away the simple pleasures that gardening should always provide.

But this book is a map that will keep you on the straight and narrow. As well as that, I believe that it will equip you to go off-route knowing that the way back home is to hand if you start to feel lost.

It is also a guide to the rhythm and connection of the seasons, which are one of the most important and beneficial aspects of gardening. It is not just a case of doing the 'right' thing according to human horticultural rules but also to the natural rhythms and laws of the seasons. Not so long ago nature was too often seen as something that the expert gardener kept firmly under control or even at bay. But you cannot garden well unless you are in tune with the rhythms of the natural world and see yourself not as its master but as a partner – and usually a very junior one at that.

Do not feel that you have missed the boat if you find yourself planting something a month later than recommended and never overlook the old maxim that the best time to prune anything is when you have the inclination and the right tool in your hand. Different parts of the country will have different timings and even within the same garden you may find that micro-climates will mean that hard and fast rules can be bent, if not broken.

But this almanac is your touchstone. It will guide you clearly and precisely when you need precision while also setting you free to do your own thing in your own time.

Above all I believe that its accumulated wisdom and experience will give you the confidence to remember the first rule of gardening: if you are enjoying it and the result pleases you, then you are doing it right.

Monty Don

JANUARY

January is a time of fresh starts. The weather is often cold or rainy and the days are still dark and short, but there is an undeniable sense of moving towards spring. Long-dormant snowdrops and aconites, as well as hellebores, catkins and, most gloriously of all, the early iris species, are all coming into flower. The days are, very gradually, getting longer. There is also a sense of urgency that grows throughout the month – especially if the weather is harsh – to get on with any jobs that need to be done in winter, before new growth appears, such as the pruning of fruit trees and the planting of bare-root shrubs, trees and fruit canes. Very slowly, the garden is starting to come alive.

Weather Watch

January can take on many different guises. This month's record highs and lows for the UK both come from roughly the same area of Aberdeenshire, with a glacial -27.2°C (in 1982) at Braemar, and a balmy 18.3°C (in 2003) just down the road in Aboyne. The potential for such extremes can make gardening a challenge! Wind direction is key in creating such extremes. From the north and east it will bring an icy blast. By contrast, south-westerly airstreams bring mild, damp weather and in this extreme example, the air is likely to have been warmed by the Foehn effect. This warming occurs when damp air is forced up over a belt of mountains and then heats up as it drops down on the other side. The wind can also be destructive. This is when the jet stream cooks up some of its fiercest storms, as clashing air masses over the Atlantic rev up into intense depressions that pack a punch on our Western shores.

- Average highest temperature in the UK: 6.4°C
- Highest number of days of air frost: 12.9, East Anglia
- Highest number of hours of sun: 63.6, the South Midlands
- Highest average regional rainfall: northern Scotland, with 21.2 days
- Area with widest range of temperatures: Braemar, with records of highs at 18.3°C and lows of -27.2°C

WEATHER ALERT

Protect your plants from wind chill. Easterly winds create some of the worst conditions for gardeners. Originating from the continent, they're not just cold, but also have very low humidity. This can dry out the foliage of evergreen plants, particularly when the soil is frozen, so protect vulnerable plants with fleece.

 ## Star Plants

10 OF THE BEST

1. Snowdrop

There is room in every garden for snowdrops: they can be tucked in every corner or even in pots and containers. Plant as bulbs in autumn or, ideally, as plants in March. Flowers January to February.

2. Witch hazel

These wonderful shrubs are subtly scented with flowers like silken tassels that appear on bare branches. Try *Hamamelis x intermedia* 'Pallida'. Flowers December to February.

3. Hellebore

The Christmas rose hellebore, *Helleborus niger*, is as pure and white as a newborn snowball. It's a real winter highlight and this one is one of the earliest. Flowers January to February.

4. Ivy

In the right place, ivy is a crackerjack of a plant. The perfect green wall: always shiny, good insulation, an excellent late source of nectar for bees and ideal winter quarters for all sorts of wildlife. Flowers October to November.

5. Dogwood

No one will miss *Cornus* on a dull winter's day. The brightly coloured, leafless stems can be fiery red, orange or vivid yellow, shining out like beacons in a bare, winter garden. Flowers May to June.

6. Winter aconite

At first, this plant looks like a tiny insignificant thing that scuttles around the base of trees and along lawn edges, but when the flowers open we're greeted with a sea of brilliant gold. Flowers January to February.

7. *Cyclamen coum*

After all the elegant skeletons, beautiful bark, urbane evergreens and modish grasses that have dominated winter, at last we come across a bit of extra colour – the valiant cyclamen pushing its pink flowers through a dusting of snow. Flowers January to March.

8. Winter heathers

Some plants are completely unfazed by winter and top of the list must be the winter heathers. Snow, ice and cutting winds leave their dainty bells shining in the weak sunshine. They tolerate all weathers and light up dark days with white, pink or reddish, bee-friendly bells. For a bright splash of colour try the pink to purple *Erica carnea* 'Vivellii'. Peak flowering from January to March.

9. Sweet box

Winter scent is much more precious than summer scent, as it seems to travel further in the cold air. This is one of the best. Make sure that you always plant them close to paths or doorways, as something so precious should never be wasted. Flowers December to March.

10. Heucheras

These useful, colourful foliage plants are easy to look after and perfect for filling gaps in the border or perking up a container display. New colours, patterns and leaf shapes appear every year, so gardeners are spoilt for choice. For drama, try 'Midnight Rose' which has near black leaves with vivid pink speckles. Flowers June to August, depending on variety.

WHY NOT TRY...

Planting up a colourful container

January is a great time to inject colour into our outdoor space, bringing a little brightness to the bleak beginning of the year. Making a large statement container is ideal for this time of year, as plants prefer not to go directly into freezing ground. Place your pot near the door or where it can easily be seen from inside the house, so you get maximum benefit. Once plants get too big for the pot, they can be removed and replanted in the ground when it warms up, to bring winter interest to your garden year after year.

'Gardening is an optimistic occupation, one that requires faith and a fundamental belief in Mother Nature.' – **Carol Klein**

PLANTS FOR A WINTER POT

- Dogwood (*Cornus sericea*) 'Flaviramea'
- Hebe 'Claret Crush'
- *Phormium tenax*
- Snowdrops
- Hellebores

PLANTS THAT CAN WITHSTAND THE COLD

These plants can withstand temperature down to -20°C
- Dogwood
- Geranium
- Asters
- Japanese anemone
- Bergenia
- Foxgloves
- *Rosa rugosa*
- Aquilegias

NOW'S THE TIME TO...

Sow perennials from seed

When we think of perennials, it's easy to picture them as potted plants, fully grown and ready to put in the ground. Gardeners often don't realise the wealth of perennials that can be grown easily from seed, meaning you are no longer restricted to what's available at your garden centre. And no more limiting yourself due to cost. You can not only produce plants easily from seed but then when it comes to planting out your newest border stars, you can be ambitious, imaginative, and adventurous! No more eking out a few costly shop-bought plants – you'll have so many to play with that new gardening vistas will open up. Plus, your plants will go on for years and themselves yield seed to make new plants.

PERENNIALS TO SOW THIS MONTH

1. **Aquilegia** – These self-sow but if you want special hybrids buy seed. Sow on the surface of good seed compost. Prick out seedlings when they've made several true leaves.
2. **Lavender** – Add grit to seed compost before sowing. Surface sow and cover with a centimetre of grit. After initial watering, don't water again until compost is dry. Prick out into gritty compost.
3. **Sea holly** – Sow thinly on surface of compost. Seed usually germinates quickly but may need a period of cold to spur it on. Prick out and pot on promptly and you may get a few flowers in the first year.
4. **Pasque flower** – Both these and clematis seeds have fluffy tails and are distributed by the wind. Push each seed into compost with its tail in the air.
5. **Astrantia** – Buy seeds, or collect from your best plants; they won't all come true but you may get some with their parents' traits. Surface sow under glass, cover with grit. If there's no germination after two months, put outside in an exposed spot.

Jobs to Do this Month

There may be less to do in the garden in January and the weather can be harsh, but there's still plenty to be getting on with. Whether you want to spend a pleasant afternoon indoors looking through seed and plant catalogues or get out and clean pots and tools or plant a bare-root shrub or tree, this is the time to get ready for the growing season ahead.

KEY TASKS

- Order seeds.
- Clean pots and tools.
- Trim off old hellebore leaves.
- Chop up and compost the Christmas tree.
- Plant bare-root shrubs and trees.
- Tie in climbers such as jasmine to prevent damage from wind.
- Plan ahead for new planting.
- Check through seed packets and throw away any that are out of date.
- Scrape off the top 5cm compost from permanently planted pots and refresh with a new layer.
- Deadhead winter bedding.
- Turn your compost to speed up decomposition.
- Check stored dahlias for rot.

GOT TEN MINUTES?

Clean your spade

Prolong the life of your spade with careful cleaning before you put it back in the shed. Doing this regularly will keep the blade in good condition. Use a coarse short-handled brush to remove all soil and debris when you've finished working with it. Wipe the handle and shaft then dry it. Once the blade of the spade is clean and dry, dip a paintbrush into some old engine oil or lubricating oil, and paint over the blade, including the cutting edge. Wipe off any excess oil with a rag then hang the spade up ready for the next time you need it.

GOT AN HOUR?

Compost your Christmas tree

Shred your Christmas tree and add it to the compost heap, where it will rot down, or use it as a weed suppressing mulch on the ground between trees and shrubs. Finely shredded woody material is useful in a compost heap as the slow release of carbon helps to power the rotting process, and keeps it from getting too slimy, particularly in closed bins. If you don't have a shredder you can hire one (if you have plenty of wood to process) or cut the side branches up small and compost those. You can saw the trunk into bits and put it in your garden waste bin or take it to the garden waste section at your local tip.

GOT A MORNING?

Protect plants from winter weather

Take steps to protect your garden from the physical damage that can be caused during a period of very cold weather. Plants are vulnerable to leaf scorch from cold winds and freezing temperatures, but protecting a plant's main growing point or crown is the priority. Good drainage and the shelter of a warm wall will also help.

- Make a temporary fleece tent around plants that are vulnerable to scorch. You can use garden twine to keep the fleece in place and pegs to attach it to the outer branches.
- Shovel snow off the paths to make them safe. If snow freezes, it can become an ice rink, so grit the paths or sweep off any patches of dusty snow that are left behind.
- Use a long cane to push heavy layers of snow off shrubs and hedges. Evergreens hold much more weight and the tops can easily snap under the strain so do those first.
- Brush snow off the greenhouse roof. A heavy layer of snow can cause damage to your greenhouse frame and it also blocks the sunlight that your plants need.

AND A FEW OTHER JOBS...

Trim old hellebore leaves

Cut all hellebore leaves right back to the ground with secateurs. Pull out any weeds too. This should leave a stand of upright flower shoots that can be better appreciated without any tatty old leaves around them. Removing the foliage now also helps to control hellebore leaf spot fungus, which causes unsightly brown scars on the leaves. New, fresh green leaves will soon emerge from the base and make a perfect low backdrop to the flowers as they continue the display well into spring.

Plant bare-root roses

Roses are supplied bare-root during the dormant winter period and need planting straight away. Prepare the site with composted manure then dig out a hole a little bigger than the root spread. Add bonemeal to enhance root establishment. Prune back the strongest roots by one-third to stimulate new growth then place in the hole. Make sure, when the soil is put back in, that the plant is not deeper than it was in the nursery. Water after firming the plant in, then mulch.

Look after your pond

Pond dwellers slow down in winter, so don't feed the fish or do anything with the dormant plants. However, they still need the water to contain plenty of oxygen, so it's a good idea to be aware of the conditions that will limit the oxygen supply. Keep pulling out fallen leaves and debris as this will use up oxygen as it rots down. Don't use any pumps or fountains when the weather gets really cold. When freezing water is agitated it freezes in deeper layers than if you leave it to freeze naturally across the surface. Leave something like a ball or polystyrene block floating on the pond so it's easy to make breathing holes for the creatures below. Sweep snow off the frozen surface, as the lack of light will stop pond-weeds from oxygenating the water.

Time to Prune

The natural reduction of energy in the garden means it's the perfect time to prune some of our beloved plants. In particular, the focus needs to be on the summer-flowering shrubs that rely on the current year's growth. It's also the last chance to do any pruning of grapevines, as they can bleed sap if it's left any later, which can weaken them.

As we are technically still in winter, now is also the perfect moment to crown lift deciduous trees, removing some of their lower branches. As they have no leaves, it's much easier to see and assess their structure.

With most woody plants dormant in winter, any pruning will remove a proportion of growth and thus channel the plant's sap and energy into fewer buds in spring. This means the resultant growth will have thicker buds and shoots, and tend to be more vigorous. So pruning is beneficial to encourage a strong framework and maintain a good shape, while keeping plants healthy and rejuvenating overgrown shrubs.

Some fruit trees can also be pruned now, before the month is up. The likes of apples, crab apples and pears can be trimmed to encourage heavy crops in the coming year, or they might be in need of complete rejuvenation.

PLANTS TO PRUNE

- Dogwoods
- Apples and pears
- Grapevines
- Buddleia
- Fruit bushes such as gooseberries and blackcurrants
- Shrub roses
- Hardy summer-flowering shrubs
- Wisteria (winter pruning)

HOW TO PRUNE WISTERIA

Before you winter prune a wisteria, identify the flowering wood or spurs. Spurs are short, budded branches which produce flower buds in the spring. Avoid cutting these out unless they have become crowded. You will often see several spurs making a little group of short stubby shoots. Look for any thin shoots winding through the main framework and cut these back to two or three buds before they create a mass of congested shoots. Cut back any sideshoots growing into the gutters or roof space as well. In some years you will need to remove one or two larger branches to encourage some fresh new growth.

In the Greenhouse

Get your greenhouse ready for the busy year ahead. You might not have many plants in there now, which gives you the opportunity to do a thorough clean. The aim is to remove overwintering pests and causes of disease, such as red spider mite and moulds.

JOBS TO DO THIS MONTH

- Choose a warm, sunny day when you can move tougher plants outside for a while. Cover them with fleece so they're not exposed to cold. Once you've swept away all debris, wash floors and benches with an environmentally friendly cleaning product. Wipe away any pools of water, then open the door and any ventilation to stop it becoming humid.
- Clear out the guttering and add the debris to the compost. Then wash through with a watering can and check the fall pipe isn't blocked.
- Check heaters are working properly. Use a max / min thermometer to measure the highest and lowest temperatures. Aim for 7°C to keep it frost free.
- Examine plants for pest and diseases. Pick off dead and damaged leaves to keep your plants healthy and looking their best.
- Wash the windows to increase the light available to plants.

DON'T FORGET TO...

- Repair fences and wooden furniture.
- Look out for pests overwintering in the greenhouse.
- Dig organic matter into veg beds.
- Open cold frames on warm days to get plants used to outdoor temperatures.

LAST CHANCE TO...

- Prune autumn-fruiting raspberries.

Fresh from the Garden

Now's the time to whittle down seed packets and throw out anything that's out of date. It may still be winter, but there are some crops you can get started, including a wide range of aubergine, chilli, and pepper varieties.

There are still harvests that can be had in January, with a little advance planning – midwinter greens such as collards and kale, along with leeks and Brussels sprouts. In the greenhouse, winter lettuce and the colourful stems of Swiss chard are a welcome sight when added to the mix of greens.

This is a good time to tidy sheds, wash plant pots, organise the seed packets and order seed potatoes. Remember to check any stored potatoes and onions for any rotting vegetables that could spoil the rest of the harvest.

WHAT TO SOW

- Sow under cover (indoors / in the greenhouse)
- Leeks
- Chillies
- Onions
- Carrots
- Radishes

WHAT TO PLANT

- Blackberries
- Gooseberries
- Plums
- Rhubarb
- Apples

WHAT TO HARVEST

- Brussels sprouts
- Kale

- Leeks
- Parsnips
- Turnips

On the Veg Patch

KEY TASKS

- Plant fruit trees and bushes if the ground isn't frozen.
- Apply lime where you plan to plant brassicas if your soil is acidic.
- Cover veg patch soil with cloche or a plastic sheet to warm the ground.
- Chit potatoes.
- Sow chillies in seed trays.
- Warm up seedbeds.
- Prune apple trees.

CROP OF THE MONTH: CELERIAC

Did you know? – Celeriac is mentioned in Homer's *Odyssey* – the Greeks called it *selinon*. Although many think of it as a root vegetable, the knobbly, bulbous bit that you eat is actually a swollen stem.

Nutrients – It is a good source of potassium, phosphorous and vitamin C and is an excellent source of vitamin K, which is good for the bones and brain.

Storing – Celeriac is best served fresh, so leave in the ground until needed. In harsh winters, it can be lifted and stored in a box of moist sand in a cool, dark place with roots still intact.

Good cultivars – 'Giant Prague' heritage variety that's good for any site or soil. 'Prinz' is reliable and less prone to bolting than other varieties. 'Monarch' has smooth bulbs for harvest in autumn or early winter.

Sow – March–April

Harvest – January–April and September–December

HERB OF THE MONTH: BAY

The leaves have a sweet, warm, aromatic scent when crushed, and are lovely in stews and soups. Leaves can be used dried or fresh. Medicinally, they aid digestion. This evergreen tree can be grown in a container when young, combining well with many herbs, or as the focal point in a sunny, warm position in the garden. It will grow to 8m tall if not pruned. Take cuttings in late summer. It can take three to six months for a cutting to strike so be patient. In colder parts of the country, move potted bay under cover for the winter.

*'At this time of year, there is no hurry. If you have not even thought about buying seed, don't worry – but it is something that is best done this month, if possible. The only crops that really benefit from sowing in January are chillies and onions, because both need as much time as possible to develop into reasonable-sized plants by spring.' – **Monty Don***

GOT FIVE MINUTES?

Test your soil

Brassicas thrive in slightly alkaline soil and potatoes in slightly acidic soil. Find out which you have by using a soil test now, while you're less busy in the garden. Then you can adapt your growing to suit.

GOT HALF AN HOUR?

Force rhubarb

Rhubarb will give you the most tender, sweet shoots in six to eight weeks if you start forcing them now. Forcing just means growing shoots in the dark. You must take the forcing pot off once you have cropped the first shoots and give the plant a nitrogen feed to help it recover and strengthen ready for next year.

Cover the crown completely by placing a forcing pot over it. You can improvise with a large dark bucket, bin or a pot tall enough to allow stems to grow to at least 40cm. Seal any holes to completely exclude the light from the plant. Stuff straw around the base if necessary, and if using upside-down pots, make sure the drainage holes are covered. Lift the cover to check the stems regularly, particularly if the days get warm as growth accelerates. If the leaves hit the top, the stems can twist and split. Remove the whole cover to harvest, gently pulling the stems away from the base by twisting out slightly. Stems should lift away from the crown rather than snap.

GOT AN HOUR?

Grow broad beans

Broad beans are an ideal crop to sow now as they revel in cooler conditions, needing just a few degrees to get started. If you live in the south with a sheltered plot and free-draining soil sow directly in the ground, but in a colder area start them off in containers somewhere sheltered or under cover and plant out in early spring. They like a sunny, sheltered position.

Beans take about two weeks to germinate. Keep them watered, and in early March, after hardening off, they should be ready to plant out. Look out for blackfly and pinch out the tips as soon as you spot them, and you should be harvesting the beans by May.

ADAM FROST'S FAVOURITE BROAD BEAN VARIETIES

- 'Aquadulce Claudia' – Hardy; suitable for early sowing
- 'Bunyard's Exhibition' – Reliable favourite with long pods
- 'De Monica' – Early cropping with high yields
- 'Masterpiece Green Longpod' – Yields large harvests
- 'The Sutton' – Dwarf variety; perfect for containers

OTHER JOBS

Protect fruit trees

The roots of fruit trees growing in the ground are insulated by soil, but tree roots in containers need protection from cold if you live in an area where freezing temperatures are common. Wrap the pots of hardy fruit trees in fleece or bubble wrap, leaving the top open.

Warm seedbeds

A few weeks before the first sowings, warm up your seedbed with a cover of clear polythene to aid germination. When the cover is removed the bed will be ready for sowing straight away. Weed the bed thoroughly then rake until it's level, with a fine tilth. Spread the clear polythene right over the bed at ground level, this keeps the soil dry and allows radiant light to warm the surface. Use heavy poles, bricks or similar to weigh it down as the wind will easily lift the cover.

DID YOU KNOW?

It's thought that iceberg lettuce got its name in California, due to the large quantities of crushed ice transported with it to keep it fresh.

TRY SOMETHING NEW... MOOLI

This winter vegetable, known as daikon radish or mooli, can be harvested in January if you sow seeds in early autumn. It has raw, crunchy heat that's perfect for a winter salad lunch. Once seedlings are big enough, move into pots in an unheated greenhouse for winter harvests.

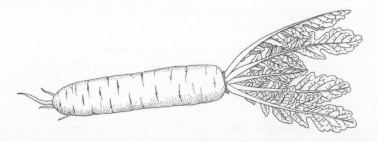

VEG TO START NOW

AUBERGINE 'LISTADA DE GANDIA'

These long, oval fruits with beautiful streaks of purple and white are productive, compact plants that are ideal for growing in containers.

How to: Sow seeds in moist seed compost, cover lightly with compost. Place seed trays in heated propagators, transplant into 7cm pots when four true leaves appear. Place seedlings under heat lamps to continue growth.

BROAD BEANS 'AQUADULCE CLAUDIA'

Slender pods filled with succulent, flavoursome white seeds. Plants grow to 1.2m tall.

How to: In an unheated greenhouse, place multipurpose, peat-free compost in Rootrainers, sowing one seed per module, at a depth of 5cm. After hardening them off, plant outdoors. Protect from birds until plants are established.

SHALLOTS 'MATADOR F1'

A hybrid producing large shallots, with firm crisp flesh. They store for several months. Delicious in casseroles.

How to: Sow 3–5 seeds per module in trays filled with seed compost. Apply base heat until germination, then move trays to an unheated greenhouse. Pot into 7cm pots once two to three true leaves appear. Plant out in spring after hardening off.

 # Rekha Mistry's Recipe of the Month

MOOLI (DAIKON RADISH) WRAPS

Makes 8 wraps

You'll need

200g raw mooli, peeled and grated
Small shallot, peeled and diced

200g potato, boiled, peeled and grated

1 tsp grated ginger

1 tsp chopped red chilli

¼ tsp turmeric powder

½ tsp garam masala

1 tbsp vegetable oil

8 tortilla wraps

Salt to taste

A few sprigs of fresh coriander to garnish

Method

1   Squeeze excess water out of the grated radish; reserve the water.

2   Heat a frying pan and add oil; sauté the shallot for a few minutes, before adding the grated potato, ginger, chilli, turmeric, garam masala and salt.

3   Add a little reserved radish water if the contents start to stick to the pan.

4   Over a low heat, allow to cook for 5 minutes before adding radish to pan. Stir to ensure mix is evenly distributed then remove pan from the heat immediately.

5   Allow to cool slightly before filling the wraps with the mixture. Wraps can be served as is or sizzled again in the pan with a little oil.

 ## Wildlife Notes

It may be cold outside, but there is plenty going on. Feeders are busy with tits and other small birds fighting for calories to keep them alive. Foxes will come in for morsels of food, and you may spot field mice hoovering up spills the birds have left behind. The days are still short and dark, although now that the winter solstice has passed, you'll notice the season turning, slowly. On mild days, you may spot bees feeding on winter clematis, honeysuckle or mahonia. Small bumblebees are the workers of winter-active colonies, which establish in autumn and use winter-flowering plants to survive the cold months. Large bumblebees are queens roused early from hibernation. If you see one on the ground, she will need nectar or sugar to give her the energy to return to hibernation. Put her on a nearby flower or, if there isn't one, mix her a 50:50 solution of sugar water and pop it in a bottle top for her to drink. She'll soon be on her way.

LOOK OUT FOR...

- Field mice foraging beneath the bird feeders. In milder regions they will continue to breed throughout winter and may need extra supplies.
- Water boatmen rising to the surface of the pond on mild days.
- Bumblebees roused early from hibernation. Grow mahonia, winter honeysuckle and winter clematis to supply them with pollen and nectar.
- Red admiral butterflies, which may fly around and bask on sunny days, before returning to their hibernation spot.
- Flies such as the bluebottle may be sunning themselves on warm, bright days.
- Slow worms, which you may accidentally dig up while preparing the ground for spring. Gently place them in your compost bin and lightly recover them.
- The long-tailed tit, which has a pale pink belly, a white head with a black stripe over the eye, and black, white and pink wing feathers. It has a longer tail than its body and makes a glorious undulating flight. You'll probably hear it before you see it – adults gather in large roving flocks of up to 20, and will call to each other continuously when moving between the trees: 'deet deet deet deet'.
- Goldcrests, which pick through trees looking for tiny morsels such as moth eggs.
- Waxwings, which visit from Europe if there are food shortages. They feed on cotoneaster and pyracantha berries.
- Fieldfares and redwings, which may still be patrolling gardens for berries and windfall apples.
- Siskins – these small finches boast striking yellow-green plumage, and have a distinct forked tail and long bill. They're most likely to come into gardens in late winter when natural sources of food are in short supply.

WILDLIFE WATCH

Top up birdbaths and break any ice to give birds a vital source of water. Listen out for the tee-cher tee-cher tee-cher of great tits. By the end of the month they'll be joined by the explosive song of the wren or, occasionally, the gentle tinkling bell of the dunnock.

DID YOU KNOW?

In January, there are only eight or so hours of daylight to forage, so a small bird such as a blue tit must feed virtually constantly. They need to eat about one third of their body weight every day just to survive.

 # Spotter's Guide to Twigs and Bark

Leaf-fall is complete now and, without their heavy summer clothes, trees offer a different, starker beauty in winter. The texture of the bark, the angularity of the twigs and the promise of the pert buds, biding their time through the winter months, all combine to define the now-skeletal silhouette. Emphasised by the cold, slanting light of a crisp January morning, a close examination of trees once admired from a distance shows how different they all are, even without their distinctively shaped leaves. Young saplings can often be hard to distinguish with their anonymous smooth barks, but by the time a trunk is as thick as your leg, it has acquired its mature texture. With some practice, trunks and outlines can be identified at a glance.

LOOK OUT FOR...

- **English oak** – The trunk's grey-brown bark achieves a thickly gnarled and deeply fissured texture with age. The brown twigs have short, stout, scaly buds becoming closer, forming a cluster at the tip.
- **Beech** – The smooth trunks have metallic or battleship-grey bark, even when fully mature. Twigs are slim, brown and slightly but

obviously hairy; slightly zigzagged with alternating brown buds that are very long, narrow and pointed outwards.

- **Sycamore** – The smooth, greenish-grey trunk flakes off in large scales to give a mottled, hoary look. The twigs are hairless and grey, with plump, speckled buds ranked in opposite pairs, with greenish outer scales.
- **Ash** – The trunk has pale-grey bark, smooth when young, but becoming furrowed as it ages. The stout twigs are pale grey, contrasting with the short, black buds. The twigs broaden and flatten at each pair of opposite buds, which face in alternating directions.
- **Hornbeam** – Bark is smooth and grey, with a network of bluish lines. Its thin, angular twigs are brown and sparsely hairy. Slim, brown buds, alternating up the stem, are tilted in towards the twig.

 ## Troubleshooting Guide

DEALING WITH FROST

In gardens, the following settings are often the most vulnerable: exposed plots, valley gardens, walled gardens (that cold air can't escape from) and plots on a slope with no exit for dropping cold air. In extreme situations, UK frosts can strike from September to June. Protecting plants from damage is relatively straightforward. Below-ground plants such as dahlias and bulbs can be lifted out of the ground over winter or protected with a deep, dry mulch. Vulnerable evergreens and shrubs can be wrapped in fleece (with good ventilation) or boxed into homemade polycarbonate frames.

WORMCASTS ON LAWNS

Removing all fallen leaves and grass clippings whenever you mow, not using organic fertilisers or composts on the lawn, and staying off the grass in winter to relieve compaction will all help reduce the number of casts. If the casts are dry, then just brush the soil into the surface of the grass with a brush. Although they can be annoying on lawns, wormcasts are valuable as they contain up to 40 per cent more humus than the soil in which the worms live and are rich in nitrogen, phosphates and potassium.

> **DID YOU KNOW?**
>
> Burrowing through the earth, worms create tunnels, which aerate and drain the soil. This takes surprising strength, and it's said that a young worm can push 500 times its own weight. Worms also digest and excrete soil, causing soil nutrients to be 'mineralised', a state in which plants can absorb them through their roots. Many earthworms also transport organic matter, such as fallen leaves, from the soil surface, down into the subsoil, increasing fertility in the root zone. Worms in turn are food for birds, snakes, mammals and insects, so they're an important element in the wildlife food chain, too.

LATE SPRING BULBS

Finding you haven't planted your bulbs by January is a fairly common situation for many people, but it's not too late. You have two choices: plant them now into containers and, once in growth, place them in your borders among other plants to display them; or, if you can see where many of your dormant plants are, plant the bulbs straight into the garden.

FROST-DAMAGED CAMELLIAS

Temperatures can fluctuate widely at this time of year, with mild conditions coaxing early-flowering camellias into flower. If this is followed by a period of cold, frosty weather the delicate petals can be damaged. The damage occurs where sun hits the frozen blooms first thing in the morning, turning them brown. So, give them protection in frosty weather by covering opening flowers with a piece of horticultural fleece. Plant new camellias against a west-facing wall or fence to avoid the morning sun.

COLD GREENHOUSES

Use a few buckets of water to keep the temperature up in an unheated greenhouse and to protect any borderline hardy plants. Placed in the greenhouse, these will gradually warm up during the day to act as a simple form of 'storage heater', releasing their warmth overnight. This will reduce the chance of a frost forming inside by providing as much as 5°C of protection in a greenhouse that has been insulated with bubble wrap.

FEBRUARY

For a gardener, February is an exciting time. The combination of the lengthening days, the increased birdsong, the burgeoning of so many spring flowers and the jobs that must be done makes it a time of business and hope. There is much to be done and much to delight in. Snowdrops, iris, crocus, the first daffodils, hellebores, camellias, hamamelis and many others are flowering. There are seeds to be sown indoors and pruning and planting to be done, too.

The weather can be cold, but the first inklings of spring are undeniable. It's a good time to fill gaps in the spring borders as these reveal themselves during the month. Buying oriental hybrid hellebores in flower, so you can choose really good colours, refreshes the borders that have a tendency to become a little muddy through their inevitable hybridisation. Adding a few euphorbias, pulmonarias, primulas, tellimas and heucheras will boost this rejuvenation, while keeping entirely in the spirit of those plants already there.

Weather Watch

The days are starting to get noticeably longer. We gain nearly two hours of daylight between the beginning and end of February. But beware. There is an old saying that goes, 'As the days grow longer, the cold grows stronger.' A glance at the figures shows that minimum temperatures in February are slightly lower than January's on average. The North Atlantic is now at its coldest so our cosy Gulf Stream blanket has worn thin and penetrating frosts can get through. Meanwhile, biting easterly winds blowing in from Siberia will act like a desiccating freeze-drier on both plants and gardeners' skin.

WEATHER FACTS

- Average number of hours of sun: Bedfordshire, 83; Northwest Highlands, 41
- Average maximum temperature: 6.6°C
- Highest number of days of air frost: 13.5, East Anglia
- Highest number of days of rainfall: 18.3, northern Scotland
- Lowest number of days of rainfall: 8.5, London and Yorkshire

WEATHER ALERT

Utilise the sunshine. Despite the cold, you can still get some benefit from the fact that the sun is climbing higher and getting a little stronger each day. Gather some of those rays and trap their warmth by putting cloches over beds where early vegetable crops will be sown.

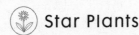 Star Plants

10 OF THE BEST

1. Early miniature irises

Reticulata irises are a popular group of small, early-flowering iris species with short flower stems. They include *Iris histrioides* 'George', which has rich purple, scented flowers and is good in pots or free-draining borders, and *Iris reticulata* 'Katharine Hodgkin'. This one looks as if someone has flicked a fountain pen at the flowers and dotted every petal with ink spots. Flowers January to March.

2. Violas

Where would our winter windowboxes and containers be without the ever-reliable viola? They fill the gap between late-autumn perennials and the first of the spring bulbs, flowering happily and surviving whatever snows, frosts and deluges that winter may send. There are many different varieties, from white to blues and purples. Try *Viola* 'Sorbet Morpho'. Flowers October to May depending on variety.

3. *Daphne mezereum* / February Daphne

Known as February Daphne because of its early-flowering nature. One of those winter shrubs like *Sarcococca* that must never be tucked away at the back of the garden. This is because, as well as being delightful, the flowers have an extraordinary scent. Plant it near a path or door that you use every day. The flowers are like pink trumpets at the gates of spring. Flowers February to March.

4. *Cornus mas* / Cornelian cherry

This is a very good tree for a middle-sized garden: ace fruit and interesting leaves in the autumn, but it is for its winter flowers that we really love this plant. Imagine, if you would, waking up on a cold February morning, flinging open your bedroom curtains to be confronted by this *Cornus*. It looks as if its leafless branches have been miraculously invaded by a profusion of one-day-old chicks. A great luxuriance of fluffy, yellow flowers that cannot help but bring cheer to the world – and this month we deserve every bit of cheer we can get. Flowers January to March.

5. *Polystichum setiferum* / Soft shield fern

If you invest in this plant (or any fern, to be honest) expecting a crowd of headily scented flowers of rainbow hues, sweet fruit or fiery autumn colour you'll be disappointed. But, if what you need is a large evergreen fern of poise and elegance, look no further. This is a really good doer.

6. Camellias

Lovely flowers (and gleamingly glossy leaves) that look amazing in a simple vase. Camellias are popular evergreen shrubs that produce a beautiful display of flowers during late winter and early spring, when little else is in flower in the garden. There are thousands of varieties to choose from, with white, pink or red flowers, which can either be single or double. Their shiny, evergreen leaves look great all year round. It does best in acidic soil. Flowers January to April.

7. *Viburnum tinus*

If you need a long-flowering, low-maintenance shrub and you have space for a plant that will get quite big, this is the one for you. It's a dense, evergreen shrub with dark and oval, glossy leaves, which contrast beautifully with fragrant pinkish-white flowers, followed by dark black fruit. It's a popular choice for hedging but may also be grown as a specimen shrub in a mixed border. Flowers February to April.

8. Daffodils

Choose some early-flowering varieties of narcissus like 'Rijnveld's Early Sensation' or 'February Gold' to get some colour into borders and pots. Flowers February to March.

9. Winter honeysuckle

Flowers are invaluable at this time of year and those that carry scent are extra special – plant a winter-flowering honeysuckle such as *Lonicera x purpusii* or *Lonicera fragrantissima* to enjoy a rich fragrance on a winter's day. Flowers December to March.

10. *Erysimum* 'Bowles Mauve'

A fantastic plant that flowers on and on from February onwards. It's an easy source of vibrant colour with spires of purple flowers above attractive grey-green foliage. Flowers February to July.

TOP TIP

Rather than discarding forced hyacinths after they have flowered, plant them outdoors to bloom in future years. Choose a sunny spot at the front of a border and add a handful of grit to the planting hole to improve drainage. Allow the foliage to die back naturally, as it will feed the bulb and give it the energy to flower well next spring.

WHY NOT TRY ...

Monty Don's favourite early-flowering clematises

- *C. montana* 'Elizabeth'
- *C. armandii* 'Apple Blossom'
- *C. alpine* 'Frances Rivis'
- *C. cirrhosa* 'Wisley Cream'
- *C. x cartmanii* 'Avalanche'

'Clematis are memorable as well as lovely and should have a place in each and every garden.' – **Monty Don**

NOW'S THE TIME TO...

Plant dahlia tubers

Alan Titchmarsh's advice: With dahlia tubers, you have three choices. First, you can pot up the tubers now and take cuttings from the vigorous new growth. Alternatively, you can pot up the tubers in early April and simply transplant them directly outdoors in late May or June. Finally, you can wait until after the last frost and plant the tubers directly outdoors. Keep the tubers in a dark, cool but frost-free place until you're ready to plant them.

The first option is easier than you think, and you'll get an extra half a dozen plants – plus you can still plant out the original tuber. Ideally you need a heated propagator for the cuttings, as well as plenty of space indoors to look after these vigorous plants for the next few months.

Before potting up your tubers, soak them in water overnight to rehydrate them. Then plant into pots of peat-free, multipurpose compost, and keep them at a temperature of 18–21°C. Leave the old stalk sticking out of the top of the compost and aim to keep the compost moist but not soggy. When the shoots are 8cm tall, cut them away from the parent tuber, ideally with a little bit of tuber attached. Dib three cuttings around the edge of a 10cm pot of cuttings compost, water them in and sit them in a heated propagator with the lid on. Keep the temperature at 18–21°C and pinch out the shoot of the cuttings when they extend beyond 10cm.

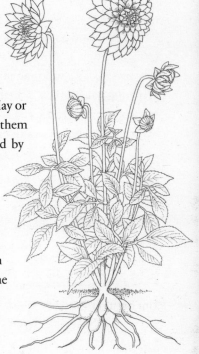

The young plants can be potted up once they are well rooted, into their own 10cm pots, then planted outside in late May or early June. But don't rush it – planting them outside too early risks them being killed by late frosts.

When planting out, sit them in the planting hole so the compost surface is fractionally below the surface of the border soil. Insert a sturdy stake to support tall varieties, as the plants can become top heavy. Tie the stems to the stake as they grow and as soon as they start to flower, start cutting blooms for the house.

DAHLIAS TO TRY

- **'Biddenham Strawberry'** – A small-flowered decorative dahlia. The rich pink petals are yellow at the base. *Height* 1.5m

- **'Primrose Diane'** – Small decorative type with lovely, pale yellow flowers. *H* 100cm

- **'Bora Bora'** – Vividly colourful cactus dahlia, that produces its large flowers for months. *H* 100cm

- **'Gay Princess'** – A decorative type with bright pink flowers. The petal tips have three points, creating a frilly look. *H* 1.2m

- **'David Howard'** – Miniature decorative dahlia with soft orange flowers and dark purple-bronze foliage. Copes well with lighter soil. *H* 100cm

- **'Kenora Valentine'** – A giant decorative type with brilliant red flowers and pointed petals. *H* 1.5m

Jobs to Do this Month

Finally, we can get out into the garden and start getting ready for spring. It's time to finish the winter pruning and begin sowing for the new season. It's a welcome mix of tidying up borders after winter and preparing your garden. Whether you're sowing sweet peas or planting bare-root shrubs and trees; sowing veg under cover or planting raspberry canes, there's a wealth of jobs to get on with.

KEY TASKS

- Start sowing seeds indoors.
- Apply organic fertiliser to borders.
- Prune late-summer clematis.
- Prune roses.
- Add a layer of mulch to borders.
- Deadhead winter pansies to keep them flowering.
- Put out food for the birds.
- Monitor greenhouse temperatures to check heaters are working.
- Turn your compost.
- Add gravel mulch to pots.

GOT TEN MINUTES?

Sow sweet pea seeds

Sow sweet pea seeds now, to plant out in borders this spring and fill the garden with summer scent or the house with cut flowers. Fill some pots with peat-free, multipurpose compost, firm them well then sow one sweet pea seed in the centre of each pot. Cover the seed with a 1cm layer of compost then soak from above with a watering can. Mice love the seeds of the legume family (peas and beans), so place the seeds in a cold frame or sheltered spot, protected by some small gauge wire mesh. They'll germinate and grow into strong seedlings before you plant them out by their supports later in the spring.

GOT AN HOUR?

Divide snowdrops

Split clumps of snowdrops as soon as they finish flowering, and replant around the garden in lightly shaded spots. They'll settle into their new homes quickly, before dying back and going dormant over summer.

Cut down into the soil around a clump of snowdrops with a spade, then lift out with all the roots.

Pull the clump apart into smaller handfuls, keeping some soil around the bulbs if you can.

Replant in a new site, at the same depth as before. Put some grit around the new clump, so you know where it is after it dies back.

GOT A MORNING?

Take root cuttings

Take root cuttings from pot-grown plants like artemisia, echinacea, acanthus and anchusa. The plant looks dormant, but cuttings will grow quickly. Expect to pot new plants on by midsummer.

1 Knock the plant out of its pot and cut through the mass of roots. Select the thickest roots from the outside of the rootball.
2 Trim each cutting to 3cm long and lay them on or insert them into some firm, gritty propagation compost.
3 Sprinkle a layer of fine grit over the cuttings and soak with a watering can. Place somewhere sheltered from rain and frost.

AND A FEW OTHER JOBS...

Trim back perennials

Trim back any remaining dead stems of perennials, to make way for the emerging new shoots. You can also cut off any tatty leaves left on evergreen perennials, before the new season's growth starts. Put all the debris on the compost heap. Tread carefully if your borders are full of emerging spring bulbs. This tidy-up will attract birds, keen to see what food you've uncovered for them. They'll eat pest larvae and slug eggs that have overwintered in the soil.

Top dress permanent containers

Replenish the compost at the top of permanently planted containers. This gives the plant a nutrient boost and encourages the fresh root development required to keep the plant healthy. Scrape off any gravel or bark mulch, then use the point of a trowel to gently loosen the top layer of compost. Be careful not to damage roots at the surface. Take out the loose compost, then add fresh John Innes potting compost. Firm it in with your fingers and water with a fine rose on the watering can. Then put some fresh mulch, of gravel or bark, over the surface to help stop weeds taking advantage of the new compost.

TOP TIP

Perennials bought by mail order now are often bare-root. Unpack them straight away, to prevent rotting. If the soil where you want to plant them is frozen or not prepared, plant temporarily in a pot of compost and keep cool.

FRANCES TOPHILL'S PLANT RECIPE FOR A FEBRUARY POT

- Lenten rose with a pink flush x 2
- Common polypody x 2
- Japanese lace fern x 1
- Australian fuchsia (*Correa* 'Marian's Marvel') x 1 (If you're in a very cold spot, a Daphne could replace this)

'I'm going to say something controversial – I like February. I know this puts me in the minority, but if you look closely there are hints of the coming spring everywhere, from bulging buds to emerging bulbs. The frustration for gardeners is that there's little to actually do outdoors. The ground is still too cold to plant much out. So why not create a container?' – **Frances Tophill**

Time to Prune

Plants strewn with spider-web garlands and the sparkle of hoar frost embody the garden at this time of year. It's biting cold and damp, making February a challenging month to work outdoors. But on a sunny crisp morning, there's still plenty to be done to keep your plot tip-top. Pruning at this time of year gives plants their form for the rest of the season and ensures the highest yield of flowers or fruit. The window for pruning apples, pears and outdoor vines is nearly over, but a new set of plants are ready to be pruned. Top of the list are group 3 clematis – get them pruned now as their new shoots are about to appear. Other key plants are hybrid tea, floribunda and English roses which can either receive their annual prune or have a revival prune for a complete rejuvenation.

PLANTS TO PRUNE

- Group 3 clematis (late-summer flowering)
- *Hydrangea paniculata*
- Hardy fuchsia
- Leycesteria
- Lavatera

HOW TO PRUNE SHRUB ROSES

Prune shrub roses before they burst into growth. Have loppers on hand, but secateurs will be sufficient for most of the work. Make clean cuts and tidy up all the prunings to reduce the risk of fungal disease. This spring pruning is particularly important for shrub roses that were left uncut for late flowers or rose hips over winter.

1. Identify the shoots crowding into the centre and cut them back flush to a main stem. This opens up the plant and improves growth by allowing air and light into the centre.
2. Cut out crossing branches by selecting and pruning out the weakest of the two or the one that compromises the shape the most. Keep an eye on the overall balance of the shrub to avoid it becoming lopsided.
3. Reduce the height by up to a third by cutting back all the whippy shoots, left from last year's late-summer growth, back to a strong bud.
4. Cut out all the remaining die-back and misplaced shoots to finish the spring tidy-up with an open, strong structure that will support plenty of flowers this summer.

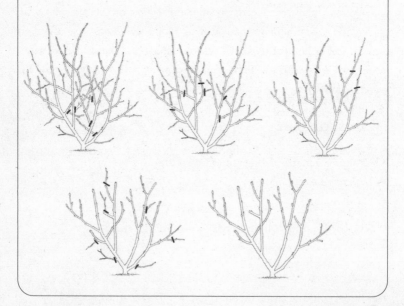

In the Greenhouse

There's plenty to do in the greenhouse this month, as the season for seed sowing gets into gear. It's time to check all the plants you already have growing undercover for pests and fungal diseases as well as start sowing crops like tomatoes. Half-hardy flowers, such as geraniums and dahlias for flowerbeds, can be sown now, as can veg crops like tomatoes that take a while to germinate and grow. Wait before you sow the fast growers like courgettes and beans – they'll get too big before you can plant them out. Ventilate the greenhouse on warm days to reduce the risk of fungal disease affecting plants.

JOBS TO DO THIS MONTH

- Sow tomatoes in small pots to get an early crop. They need 22°C in a propagator if possible for reliable germination.
- Prune greenhouse climbers such as passion flowers and bougainvillea.
- Check plants regularly to reduce pests and fungal diseases in your greenhouse. Snip off yellowing leaves before they start rotting and look out for aphid on new buds or mottled leaves that indicate the start of red spite mite feeding.
- Prick out seedlings grown in small pots, once they've developed two true leaves.
- Spray fuchsias with water on warmer days to encourage new growth.

DON'T FORGET TO...

- Plant spring bedding in pots or near to the house if the weather is mild.
- Look out for weeds. As the weather warms up, weeds can quickly cover a border. Keep on top of weeding to save time later and ensure that emerging plants can develop without competition for nutrients, light and water.
- Sow hardy annuals in modules or pots in peat-free compost.

LAST CHANCE TO...

- Prune apple and pears before the new buds begin to open.

Fresh from the Garden

This is the time that seed sowing can begin in earnest, starting with the hardier vegetables such as beetroot, spinach and winter lettuce. You can sow these in plugs and seed trays, so they can germinate and grow into strong seedlings in the protection of the greenhouse. They can then be hardened off and planted outside when the soil warms up in March or April. At this time of year, you can still be harvesting the last of the parsnips and leeks, and enjoying the tender stems of purple sprouting broccoli.

WHAT TO SOW

- Broad beans
- Chillies
- Leeks
- Peas
- Tomatoes

WHAT TO PLANT

- Rhubarb
- Quince
- Apple
- Pear
- Bare-root raspberry canes

WHAT TO HARVEST

- Brussels sprouts
- Forced rhubarb
- Kale
- Purple sprouting broccoli
- Leeks

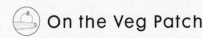 On the Veg Patch

KEY TASKS

- Start sowing indoors.
- Prepare beds for sowing.
- Plant out shallot sets.
- Check stored fruit and remove any that are showing signs of rot.

CROP OF THE MONTH: JERUSALEM ARTICHOKE

A sweet and nutty alternative to potato, grown for its tuber and related to the sunflower.

Did you know? – Jerusalem artichokes produce yellow flowers that are a source of nectar for bees and other beneficial insects.

Nutrients – Jerusalem artichokes are a good source of potassium and iron. They also have a prebiotic effect, helping promote a healthy digestive system.

Storing – Tubers don't store well and are best left in the ground until required. They can be kept for a short while in a paper bag in the fridge.

Good cultivars – 'Fuseau' produces long tubers with a smooth skin, which makes them easier to peel.

Plant – February–May

Harvest – December–February

HERB OF THE MONTH: ROSEMARY

Rosemary is the most brilliant herb, as the leaves can be used all year round in stews, soups, roast vegetables, roast meat and on the barbeque. Rosemary helps to digest fatty foods. This evergreen herb makes an attractive container plant, either on its own or as a centre piece. Take cuttings in late spring, from non-flowering shoots. This is by far the fastest method of propagating rosemary – otherwise, sow seed from late winter to early summer.

GOT FIVE MINUTES?

Boost growth

Apply organic fertilisers, such as manure or chicken pellets, to your soil now. They are slow to release nutrients, so will last in the soil for the whole life of many crops. A mulch of well-rotted manure will improve soil structure and give a good long-term slow release of nutrients. For a faster spring feed, apply blood, fish and bone, chicken pellets or organic granular fertiliser before sowing or planting.

GOT HALF AN HOUR?

Plant shallots

Shallots like a sunny spot with light, well-drained soil. Dig in lots of garden compost before planting, and rake in some fertiliser too. They also need plenty of room to develop their clump of bulbs, so space the sets further apart than onions, to ensure a large crop. Harvest the bulbs for storing when the foliage turns yellow.

- Make a straight, shallow drill using a trowel or hoe. Space the sets 20cm apart, with 30cm between rows.
- Push the soil back into the drill around the sets, leaving the papery tip just above the surface. Firm gently in place.
- Water along the row with a fine hose spray or a watering can with a rose. If any sets get dislodged, replace carefully.

GOT AN HOUR?

How to chit potatoes

We chit potatoes because if they are given a head start and planted with young shoots that are already growing they are likely to produce a crop sooner than if planted totally dormant. This means that you can be enjoying your new potatoes earlier than if they had not sprouted at all. It's very easy and once the shoots are 2–3cm long the potatoes can be planted out, safe in the knowledge that growth is under way.

1 Don't bother chitting main crop varieties, which stay in the ground longer. However, you must chit first and second early varieties (the quickest to mature).

2 Examine each tuber – you'll find more tiny buds, or eyes, on one end. Stand tubers in a bright, frost-free place, eyes up, to sprout. An egg box will keep them upright

3 Start them off now as it will take several weeks for shoots to grow. Don't discard any with longer shoots (the result of being in light too early or in dark too long).

4 Rub off excess shoots, leaving three or four per tuber – they should be about an inch long. Leave too many shoots per tuber and you'll get a lot of tiny spuds on each plant.

'Many folk don't have room for masses of potatoes on their veg patch, but a short row of first earlies can be fitted into most gardens to produce tasty new potatoes. The earliest crops come from seed potatoes (raised specifically to be healthy and virus free) that are chitted before being planted. The word chit dates from the 17th century and means "a shoot", so chitting is the act of encouraging seed potatoes to sprout before they are planted.' – **Alan Titchmarsh**

ALAN'S FAVOURITE EARLY POTATOES

- **Duke of York** – Can also be grown as a main crop variety
- **Foremost** – Firm and waxy, so good as new potatoes
- **International Kidney** – The Jersey Royal potato
- **Pentland Javelin** – Disease resistant and tasty in salads
- **Rocket** – A fast grower, can be grown early under cover

OTHER JOBS

Tidy herbs

Neaten up your herb bed before new shoots appear. Snip off any remaining seed heads from herbs such as marjoram and fennel. Then weed around them and aerate the soil surface. You can also tidy up any unruly shoots of woody herbs such as sage, thyme and rosemary, but don't prune hard yet. Then add a layer of garden compost around vigorous green herbs like fennel, but avoid enriching the soil around evergreen grey-leaved types. Spread grit around these instead, to keep weeds at bay.

Protect crops

Watch weather forecasts carefully and be ready to protect the emerging shoots of broad beans and autumn-planted onions and shallots from the hardest frosts. Use fleece, and make sure it's weighted down, as the slightest gust of wind will carry it off. Keep brassicas netted to protect them from pigeon damage. But if it snows, knock the snow off before the weight tears the netting. Cover seedbeds with polythene to warm the soil for early sowings. It also protects the ground from waterlogging after heavy rain.

Get veg sowing under way

You can begin sowing tender crops indoors and hardy ones outdoors now. Start off tomatoes, broccoli, salads and globe artichokes in a heated propagator, if you have one. You can also sow Brussels sprouts, radishes, leeks, onions, peas and spinach under glass. In warm areas, sow hardy peas and lettuces outdoors. But keep fleece handy, in case bad weather is forecast, to protect young seedlings.

TRY SOMETHING NEW...

Alfafa sprouts

These crunchy shoots are a protein powerhouse and are easy to grow indoors all year round. Use a sprouting jar or make holes in a jam-jar lid. Cover the bottom of the jar with seeds, but don't overload, then fill with water and soak for 12 hours. Drain and lay the jar on its side, rinsing and draining twice a day until the seeds have sprouted. Use in salads for a nutty crunch. Sow or sprout at any time of year. Buy seeds suitable for home sprouting.

 ## Rekha Mistry's Recipe of the Month

KALE BHAJIYA

Makes 15 generous spicy batter balls

You'll need

1 cup gram (chickpea) flour

¼ tsp turmeric powder

1 tsp caraway seeds (crushed)

½ tsp fennel seeds (crushed)

½ tsp garam masala

1 tsp dried coriander leaf

1 tbsp grated ginger

1 tsp finely chopped green or red chilli

1 tbsp vegetable oil, plus a sufficient amount for frying

1 tsp fine sea salt

water to bind

6–7 kale leaves, deveined and roughly chopped

1 medium onion, peeled and sliced

1 tsp bicarbonate of soda

Method

1 Place the gram flour in a large bowl and add the spices, chilli, oil and salt, rubbing into the flour.

2 Add a little water to the flour until it forms a thick 'dropping' batter.

3 Allow the batter to marinate for an hour.

4 Heat a deep-fat fryer or 5cm of oil in a deep pan. A drop of batter will sizzle when the oil is hot enough.

5 Add the kale, onion and bicarbonate of soda to the batter and mix well.

6 Carefully drop a tablespoonful of the kale batter into the oil and fry until the dough balls are golden brown and cooked through.

7 Serve hot or cold with a spicy tomato chutney.

Wildlife Notes

February brings a glimmer of hope: snowdrops and the first of the daffodils dare to bloom, birds sing louder, some bumblebees wake up, but there may be prolonged freezing and snow just when you'd started to breathe a sigh of relief and look ahead to spring. Continue leaving food out for birds, which are trying to get into shape for breeding, and start leaving cat biscuits and water for hedgehogs, too. Don't worry about frogspawn freezing in your pond – any below the water's surface should survive an icy spell. Bees and other pollinators rely on nectar-rich blossom and will suffer without it.

Robins, blackbirds and thrushes are now on the last of the autumn berries – from pyracantha, holly and ivy. Leave halved apples and grated cheese on the ground for birds and think about using what's left of the bare-root planting season to plant a berrying shrub – it'll offer birds winter food for years. We're well into breeding season for foxes now; you may hear the shriek of the female at night, followed by the 'hup-hup-hup' of the male. Soon, their cubs will be playing on the lawn. We may yet see the worst of winter, but spring is on the breeze.

LOOK OUT FOR...

- Hoverflies – early-flying types, including the marmalade hoverfly, *Episyrphus balteatus*, are tentatively on the wing now.
- Blackbirds. A sign of spring, they may start to sing if it's mild this month.
- Robins engaging in 'courtship feeding'. Breeding pairs are getting together now and will be nesting soon.
- Queen bumblebees, which fly low in zigzags, looking for somewhere to start a nest.

- Red Admiral butterflies – on sunny days, you may find them basking on a sunny wall or fence, or investigating buildings, looking for a new place to rest.
- Yellow brimstone butterflies may appear on sunny days.

WILDLIFE WATCH

Start leaving out water, and meat-based dog or cat food for hedgehogs rising from hibernation. They will be particularly hungry after having gone several months without food. Leave the food out from dusk and cover or discard any that is left first thing in the morning to prevent flies from laying their eggs in it.

DID YOU KNOW?

Found mainly in garden ponds, the great water boatman swims on its back, just below the surface of the water. Adult backswimmers can fly, and the motion of its back legs resembles the oars of a boat. A bite from one can be painful, as its saliva is toxic. It's not to be confused with the lesser water boatman (*Corixa punctata*), which is smaller and swims on its front.

WILDLIFE PROJECT: HOW TO MAKE A BUTTERFLY FEEDER

Butterflies need nectar to give them the energy to fly and find a mate. We should aim to grow open, single flowers from March to November, but sometimes there may be no flowers for butterflies to drink from. This simple feeder, using a sugar-water mix, will give them a helping hand.

You will need:

Sugar

Water

Saucepan

Clean sponge

Bowl

1. Add roughly equal amounts of sugar and water into a saucepan.
2. Boil gently, stirring constantly with a wooden spoon, until the sugar has melted.
3. Cut the sponge to size and place it in the bowl. It should fit snugly.
4. Gently pour the sugar solution onto the sponge, so it will be absorbed.
5. Place outside so butterflies and other pollinators can take a drink.

Spotter's Guide to Crawling Bugs

It is deepest winter and you won't find many insects flying about in your garden, but there are still plenty of crawling things that are surviving under rocks and logs. Here, in frost-free pockets, these invertebrates can eke out a slow hibernation existence, rousing briefly when the temperature rises, but rarely leaving the safe shelter of their winter roosts.

Predator and prey often share the same insulated space at this time of the year. They don't move for days (or even weeks) as they seek refuge from the greater danger – freezing cold weather and insidious wet.

If you disturb any of these creatures when you're tidying up your garden, scoop them up gently and release them in a rough corner where an alternative log or rock refuge is available.

- **Common woodlouse** – Broad, oval, shiny, brownish-grey, often flecked with large white or pale yellow dashes. Gently flanged around the edge so that it can clamp down and conceal its 14 legs and antennae when disturbed.
- **Centipede** – Broad, flat-bodied, with 15 pairs of long legs (one pair to a segment) giving this predator a good turn of speed. An extra pair of 'legs' has developed into curved, venomous fangs, though these are too feeble to puncture human skin.
- **European earwig** – Dark brown with pale wing-cases tipped with yellowish flaps. The tail pincers are deeply rounded in males, less curved in females. Waves its forceps threateningly but they are harmless to humans.
- **Pill millipede** – Smooth, deep chocolate-brown to charcoal-grey and shiny, but the side and hind edges of its segments have pale margins. Very convex, rolls into a tight ball, concealing its 15 pairs of legs and antennae. Its tail has a single large convex segment.
- **Snake millipede** – Smooth and cylindrical with 80–120 pairs of legs (2 pairs per segment), allowing it to push, worm-like, through loose soil. It coils into a tight spiral, legs inwards, while hibernating or if it is disturbed.

 # Troubleshooting Guide

This is an ideal time to add organic fertiliser to the soil around the base of hungry plants. Your targets should be any that were shy to flower last year or showed signs of nutrient deficiency (such as yellowed leaves), as well as fruit and veg crops. Most garden plants don't need feeding every year, so save your attention and money for those that do. Organic fertilisers include pelleted poultry manure (high in nitrogen), seaweed meal (rich in trace elements) and blood, fish and bonemeal (a balanced fertiliser with plenty of phosphorus for root growth). They are not synthesised in a factory but come from plants and animals. Their natural origin means that these fertilisers take time to break down into the chemicals that plants can use. This process is done by tiny organisms in the soil and takes several months – which is why it's a good idea to apply organic fertilisers now. The nutrients will then be ready to feed your neediest plants in spring, just when they're ready to absorb and benefit from them.

DEER TROUBLE

Deer can wreak havoc on garden plants, stripping flowers and foliage and damaging tree bark. To keep all species out completely, you need a 1.8m high fence with partially buried posts and a self-closing gate to let out any that do get in. But if you just want to control their behaviour while they're in your garden, grow the plants they're less fond of and place guards around trees and shrubs. Chemical repellents have mixed reviews and may just divert them to other plants, but the scent and sound of a dog in the garden may scare them away. Also try growing plants they like, such as rosebay willowherb, dandelions and brambles, to distract them from your ornamentals. Plants they're unlikely to nibble include hellebores, rhubarb and common jasmine.

DID YOU KNOW?

Poppy seeds can still germinate after more than 100 years in the ground.

FAILED SEEDS

If sown too deep or too shallow, a seed won't have the necessary trigger – light, moisture or heat – to kick-start growth. So always follow sowing depth instructions. Temperature can also be a challenge to get right – if seeds are too hot or too cold once sown, most won't get the trigger that tells them to get a wiggle on and start germinating. Whether in a glasshouse or on a windowsill, seeds of tender plants (such as tomatoes) ideally need around 20°C to germinate, while cool climate species (such as lettuces) prefer around 10°C. After sowing seeds, stand the pot or tray in a shallow dish of water, rather than watering from above. Overhead watering can wash the seeds out of place, reducing germination rates.

NO CAMELLIA FLOWERS

To keep your camellia healthy and flowering well, make sure that the soil is adequately moist at all times, so whenever it starts to dry out, give it water, ideally rain water. In addition, make sure that it is kept well mulched with bulky organic matter, 8cm deep and applied in a circular band extending from close to the trunk to just past the outermost spread of the branches. Avoid using any nitrogen-rich materials for the mulch as excess nitrogen is often responsible for plants showing great growth but very poor flowering. Finally, apply an ericaceous feed from early spring until early June.

BARE FENCES

A fail-safe choice to cover boundaries is fragrant-flowered star jasmine (*Trachelospermum jasminoides*). Most say it likes full sun, but it's pretty resilient and can cope with part shade, and pretty much any type of soil. It may have a few less flowers, but will still get big enough for your trellis-topped fence, eventually up to 10m. Another classic evergreen climber is the fragrant, pink-tinged *Clematis* 'Apple Blossom'. Plant either of these in spring or autumn, and no pruning is necessary apart from to shape them or keep them within bounds. For something a bit different, try *Holboellia brachyandra*, a woody twining evergreen climber that needs a sheltered spot and frost protection. In March it bears highly scented white to pale-blush flowers. If pollinated, these are followed by purple sausage-shaped edible fruit.

OVERGROWN VIBURNUM

Viburnum tinus is a very robust and vigorous shrub that can cope with being pruned to ground level and will regrow readily. Prune it in late spring, once there is no danger of frost and before birds start nesting.

Work out what height you would like it to be in the long term, then cut back to at least 30cm below that, to allow for strong regrowth. Pruning in spring also means you won't have to look at a bare shrub and pruning cuts for too long. After pruning, feed with a general slow-release fertiliser, mulch with garden compost and water during dry periods. In future years, cut it back every spring to keep it within its allotted space.

MARCH

March is irresistible. It can be like a naughty child – throwing tantrums of snow and ice, gales and rain – often all in the same day. But just when your patience is wearing thin it charms you. This is the month when bulbs really take charge, growing and flowering regardless of the weather, adding an array and intensity of colour that sweeps winter away. Daffodils, irises, crocus, the first species tulips, muscari, chionodoxa, scillas – they are all flowering strongly and even a few snowdrops linger into the beginning of the month. The days are getting longer – the clocks change at the end of the month – and March, out of all the months in the gardener's year, is the one where time presses most urgently. Winter jobs need to be finished along with sowing, planting, getting that feel of soil on your hands and just a little sun on your back.

Weather Watch

We weather people tidy the year into three-monthly chunks, and March is the first of the meteorological spring months. It also contains the important milestone of the equinox, as the sun crosses the equator on its way back to us for summer. But don't drop your guard – snow falls as often in March as it does in February and a temperature of -22°C was recorded at the aptly named Logie Coldstone in Aberdeenshire in 1958. The notorious winter of 1947 saw blizzards sweep through to mid-month, and Norfolk farmers needed pneumatic drills to dig turnips from frozen ground!

- Equal lengths of day and night occur around 20 March (spring equinox) and again around 22 September (autumn equinox).
- Highest average regional rainfall: northern Scotland, 21.2 days
- Lowest average regional rainfall: London, 9.3 days
- Highest number of hours of sun: the south east of England, 120.5
- Highest number of days of air frost: East Anglia, 10.4
- Average highest temperature: 8.9
- Average lowest temp: 2.1

| WEATHER ALERT

A run of mild days may tempt us into reaching for the seed packets, but damp soil takes a while to warm up, especially heavier soils and those containing clay. It may be worth investing in a simple soil thermometer, to avoid early sowings rotting in cold, wet conditions.

 Star Plants

| 10 OF THE BEST

1. *Chaenomeles*

This is the Japanese quince: a spiny shrub that sweeps into the gloom of winter with all the verve and ebullience of a wet puppy playing in a pile of laundry. At this time of year, what we all need is some zing and this plant does not disappoint. There are a few varieties, mostly in reds and pinks that all flower early and then have delicious smelling, yellowy fruits in autumn. Flowers March to May.

2. Primroses

The primrose is the iconic spring wildflower that grows across Britain and Ireland – everyone knows and loves it. Hedgerows, woodland, north-facing banks – the primrose thrives anywhere that's shady and moist. It's by far the most widespread of our five native primula species and, along with the cowslip, is a parent of the garden polyanthus. Try *Primula vulgaris*, the classic pale-yellow blooms on arching stems. Flowers January to April depending on variety.

3. Pulmonarias

Pulmonaria is commonly known as lungwort, so I suppose it could draw attention to the importance of breath control while singing? No? OK, then let's just talk about how useful this plant is as a bit of shady ground cover. It will happily spread around and bring joy to the tricky areas under trees. It also provides very useful early food for bees. Try 'Blue Ensign' or *Prunus angustifolia*. Flowers March to May.

4. *Amelanchier lamarkii*

This is one of the very best trees for a small garden, as it gives three rousing crescendos. First these amazing flowers in spring, as the young leaves emerge, followed by edible berries in summer, then, to finish off, spectacular autumn foliage colour. Flowers March.

5. Pussy willow (*Salix caprea*)

This is pussy willow, although I have always thought that the fluffy buds were more like the ears of rabbits that anything feline. Just to confuse the zoology even more, it is also known as goat willow because it was used as goat food. However, if we ignore the larger mammals this is a fantastic plant for bees of all shapes and sizes who queue up to harvest the pollen. Male and female catkins are produced on separate plants – males are showier. Takes easily from cuttings plunged in beds or pots outside. Good coastal or wild plant. Flowers March to April.

6. *Clematis armandii*

This is one of the earliest flowering clematises. It is incurably romantic and, to make it even better, has long strappy evergreen leaves to announce its presence for the rest of the year. It's vigorous so needs to be cut back every few years. Prune immediately after flowering. Can be a bit tender in colder gardens. Flowers March.

7. Leopard's bane (*Doronicum orientale*)

At this time of year, with daffodils everywhere, you might think we don't need any more yellow. You would be wrong. This plant has yellow, daisy-like flowers that bring another timbre of smiling sunshine to our gardens. A bit more subtle than a bank of daffs. Flowers March to April.

8. Forysthia

This popular plant will light up your borders – they are reliable, easy to grow spring shrubs. They are excellent for growing against walls or as large, open shrubs in the border. Flowers February to April.

9. Spring anemones

One of the most delightful features of the spring garden is the low tapestry of early flowers that are at their peak in the shade before tree and shrub cover closes over. Hellebores and primroses, epimediums and pulmonarias, erythroniums, dwarf bulbs and more are all knitted together by the low, steadily spreading growth and upturned flowers of woodland spring anemones. These sprout mainly from brittle runners to present neat, ferny foliage topped by upturned flowers in a range of soft and bolder colours. Once most snowdrops have finished their blooming season, these anemones nudge around other spring perennials and bulbs, and intermingle with them to create an intimate embroidery of colour, to then fade away for summer. By contrast with the sun-loving Mediterranean anemones of late spring and summer, these enjoy dappled shade. Try 'Royal Blue' or the ruffled 'Vestal'. Flowers March to April.

10. Pulmonarias

These shade-loving perennials produce bee-friendly flowers that come in shades of blue, violet, pink, purple, red and white. They are also excellent plants for shade. Try 'Blue Ensign' which has striking blue flowers. Flowers March to April.

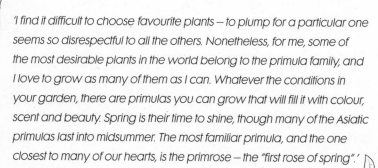

'I find it difficult to choose favourite plants – to plump for a particular one seems so disrespectful to all the others. Nonetheless, for me, some of the most desirable plants in the world belong to the primula family, and I love to grow as many of them as I can. Whatever the conditions in your garden, there are primulas you can grow that will fill it with colour, scent and beauty. Spring is their time to shine, though many of the Asiatic primulas last into midsummer. The most familiar primula, and the one closest to many of our hearts, is the primrose – the "first rose of spring".'
– Carol Klein

WHY NOT TRY...

Carol Klein's favourite primulas

- *Primula vulgaris* – The purest, most simple of all primroses which, as it is native, always looks at home. Plant in soil rich in organic matter, in a spot that is shaded for part of the day. **Flowers** Mar–May **Height x Spread** 20cm x 25cm

- *Primula bulleyana* – One of the Asiatic candelabra primulas, *P. bulleyana* has striking orange flowers held in whorls around the stems, opening from the lowest first. Loves moist, boggy soil. **F** May–Jul **H x S** 60cm x 30cm

- *Primula vulgaris* '**Lilacina Plena**' – This is one of the oldest recorded doubles, with a beauty unmatched by modern hybrids. It is sterile so must be divided to make more. **F** Mar–May **H x S** 20cm x 20cm

- *Primula vialii* – People are always intrigued by this primula with its red-hot poker style spikes. Lilac flowers open from the red buds at the base first. It enjoys damp, acidic conditions so makes a perfect bog garden subject. **F** Jun–Jul **H x S** 60cm x 30cm

- *Primula* **Gold-laced Group** – The black, velvet-like flowers are held symmetrically at the top of a straight stem. Together with the gold-edged petals this creates a formal impression. It is a type of polyanthus and so is just as easy to grow as the more common, brightly coloured types. Grow it from seed and then divide the best plants with the darkest petals and clearest edges. **F** Mar–Apr **H x S** 25cm x 30cm

Jobs to Do this Month

There's no shortage of jobs to get on with in March – this is the month where gardening takes off. This month, it's a mix of preparation for the season ahead, with tasks like digging over and mulching beds and borders, as well as getting started on sowing and planting. It's time to sow many veg and flower seeds indoors, sort out the lawn and plant up summer bulbs among many other enjoyable and exciting jobs.

KEY TASKS

- Mow the lawn in dry weather with the blades on a high cutting setting.
- Deadhead and prune hydrangeas.
- Tie in climbers.
- Move plants that are in the wrong place.
- Get early potatoes in the ground.
- Clear dead leaves from ponds.
- Deadhead faded daffodils.
- Check tree ties haven't become too tight.
- Check for aphids on new shoots.
- Check soil temperature.

TOP TIP

Make sure seed compost is not too wet or heavy. If you can squeeze water out of it before sowing, it is too wet.

GOT TEN MINUTES?

Tidy up the lawn

Brush worm casts off the lawn with a besom broom on dry days.

GOT AN HOUR?

Sow seeds

Sow seeds in a greenhouse, on a windowsill or, when the ground permits, outdoors where they are to grow. Hardy annuals such as pot marigolds (calendula) and nasturtiums are tough little plants that give colour in summer for very little effort and expense.

GOT A MORNING?

Plant a shrub

Plant pot-grown shrubs now that the weather is mild. Choose one that looks healthy and well balanced in the pot, and is weed free. Check the roots by removing the pot if you can; they should have filled the pot without circling. Place the shrub in water to soak the roots before planting and water again once it's planted.

1 Dig a hole twice as wide but the same depth as the shrub's pot, then break up the bottom of the hole with a fork, add a shovelful of compost, then water the bottom of the hole.

2 Knock the shrub out of the pot and place it in the centre of the hole. Always keep the stem base of woody plants just above the ground to stop them rotting. The compost level from the shrub's pot is usually the best guide for correct depth.

3 Backfill around the roots with soil or a mixture of soil and compost if your soil is poor. Put a bit in at a time and tread it down as you go, while holding the shrub upright and at the right depth.

4 Use secateurs to trim back any damaged stems to a healthy bud. That shoot will soon be replaced.

GOT A DAY?

Move perennials

Now is a great time to decide whether any of your hardy perennial plants would look better in a different spot and sort our your borders. Hardy perennials are the plants that tend to die back to the ground each winter and resprout in spring, ready to bloom in summer and autumn. Gardeners often cut them back in winter to get rid of their dead stems and leaves.

At this time of year, the live parts are only the roots and new buds, which means these plants won't be disturbed much if you move them now. Your soil should now be hydrated by winter rain but about to warm up with spring sunshine, so conditions are good for the plants to get established. This is a good time to move asters, rudbeckias, sedums, heleniums and Japanese anemones. They will have plenty of time in their new home to get ready for blooming in a few months' time.

You can also divide any that have got too large. Keep your perennials vigorous and flowering well by dividing clumps every few years. With fleshy rooted hostas, daylilies and red-hot pokers, it's best done in spring to avoid rot. Dig up the clump and pull it apart or cut it into several pieces with a knife. Discard the old centre, then replant or pot up the outer sections. Water regularly until well rooted.

AND A FEW OTHER JOBS...

Mow your lawn regularly

The lawn is made up of thousands of small grass plants that after a trim will thicken up their growth from the buds in the centre of the plant – so mowing is a bit like pruning. Keep the cutting blade high at the start as this avoids damaging the buds and strengthens the grass's ability to compete with lawn weeds, which are adept at tucking close to the ground and are rarely mowed off even if you lower the blades. Don't mow grassed areas with bulbs yet – wait for the bulbs to die back first. Wait for a dry day and for the grass to grow about 7cm before mowing for the first time.

Repair a lawn edge

Cut out pieces of turf at the edge of the lawn where they've become worn or dropped. Make the vertical cuts first and aim for an even shape that has good-quality grass behind the damage, with a thick layer of soil beneath. Lay the piece of turf out and turn it around so that the good edge forms the lawn edge. Pack soil under the new edge to make it level. Tamp it down and add soil to the gaps before sowing some seed to fill them. Give it a final water to bed it in.

Tidy up ponds for spring

Remove any netting that was laid over your pond in autumn to catch leaves. Amphibians returning to it to breed may be deterred or trapped if netting is still in place. Also scoop out any old plant debris. This can reduce the oxygen in the water, decomposing and producing gases that are harmful to pond life. Lift and divide any perennial plants you want more of. You can also draw up a list of new plants you want to buy in April. Check the pond water is fresh and ready for the burst of spring activity. If the water smells or there is build-up of sludge on the pond bottom, it's time to wade in for a major clean-up and water change.

Pot up lily bulbs

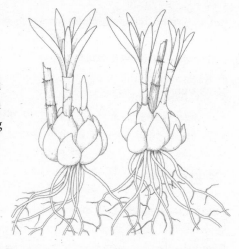

Select large deep pots for growing lilies so you can plant several bulbs in each. Part-fill the pot with rich loam-based potting compost. Place the bulbs on the surface, with the scales pointing upwards. Space them about 5cm apart, then cover with a layer of compost equiv-alent to the height of the bulb. Water well to settle the compost, then spread a layer of grit on the top to deter slugs. Lilies are hungry, so feed regularly all summer. You can also plant summer-flowering bulbs such as galtonias, the smaller varieties of gladioli and lilies. Position them in gaps between border perennials.

Improve soil in beds and borders

Dig over beds with a fork, working in a good sprinkling of blood, fish and bone meal. In wet winters much nutrition will have been washed away – the fertiliser will replenish it. Then mulch beds and borders with chipped or composted bark to seal in moisture and help keep down weeds throughout the growing season. Alternatively, use well-rotted garden compost. A feed followed by a mulch will power up your border.

'Your soil is the most important thing in your garden. Be kind to it, respect it, enrich it often and your plants will grow. Ignore it and they'll sulk. Soil is like children – it thrives on regular food and drink.'
*– **Alan Titchmarsh***

DON'T FORGET TO...

- Cut down the spent stems of hardy perennials to make way for new shoots.
- Continue to cover tender plants with fleece on frosty nights.
- Replace any pond pumps removed over winter.
- Sow hardy annuals.
- Protect young plants from slugs.
- Plant out hardy perennials and spring bedding like primulas and violas.
- Water your spring pots.
- Fork out weeds from flowerbeds.

LAST CHANCE TO...

- Finish planting bare-root trees and shrubs as they need to be in place before it gets warm. Container-grown trees and shrubs can be planted any time (as long as the ground isn't frozen).

*'Whatever the weather does, spring cannot be denied. March birdsong is the best of the year and the bulbs, from the latest snowdrops to the earliest tulips and a dozen species in between, are all bursting into flower. It is also the month when I finish all my winter pruning. Roses, buddleias, sambucus, all the late-flowering clematis, Cornus, willows, fuchsias, hydrangeas – the mopheads and lacecaps lightly, Hydrangea paniculata boldly – all can be cut back now to encourage the new growth that will bear flowers in summer. There is something of a rite of passage about this – last season's growth is removed and winter is formally over. There is no going back – spring is here.' – **Monty Don***

Time to Prune

Our gardens really spark back into life this month. Bulbs are emerging, perennials are producing fresh growth and deciduous plants are sprouting new leaves. As the days get noticeably longer and the mercury rises, our gardens and their inhabitants are warming up for the season ahead.

In colder parts of the UK, there's still time to prune apples, pears and late summer-flowering group 3 clematis, but in warmer areas it's getting too late. Climbing roses, hybrid teas and floribundas are still fine to prune, but the sooner the better, so they don't waste their energy growing leaves that will be pruned off. Borderline-tender perennials, such as penstemons and phygelius, can be pruned too.

But the key plants to prune this month are evergreens and shrubs that flower on this year's growth, such as buddleias. Check them for birds' nests first, and leave alone if nests are in use.

HOW TO PRUNE... *BUDDLEJA DAVIDII*

This popular buddleia species is best pruned in early spring, between late February and early April. Pruning before it starts into growth means it doesn't waste energy growing new stems that are destined to be removed. Buddleia is a vigorous shrub and can be cut back hard annually, 50–100cm from the base. This triggers the production of new stems that will carry lots of butterfly-friendly blooms from July onwards.

1 Cut back into the old woody framework of the buddleia, aiming to leave it about 50–100cm tall. With thick branches, make life easier by using a clean, super-sharp, sickle-shaped pruning saw. Cut at a slight angle to ensure rainwater runs off easily.

2 Make your cuts just above a dormant bud or new shoot. It's not always easy to spot dormant buds on old stems, but do your best. Prune thinner stems with sharp secateurs and be sure to cut cleanly, so you don't leave any snags or tears.

3 Tidy up the cut stems – those under 1cm thick can be chopped up and put directly onto the compost heap. Thicker stems can be piled in a quiet corner of the garden, where they will provide food and shelter for many kinds of wildlife.

Tender evergreen shrubs are best pruned in spring. Waiting until now should ensure that the new growth triggered by the pruning won't be damaged by hard frosts. Evergreen hedging and topiary can be trimmed now to create a smooth finish. With established evergreen shrubs, a light natural-looking thinning out will improve aesthetics, air circulation and light levels beneath for other plants.

PLANTS TO PRUNE

- Penstemons
- Phygelius
- Winter-flowering deciduous viburnum
- Cotinus
- Deciduous grasses
- Forsythia, after flowering
- Hebe

In the Greenhouse

Things are hotting up in the greenhouse this month. But at the same time as sowing seed and looking after plants that are already growing in the greenhouse, keep an eye out for any emerging pest problems and make sure your greenhouse is in a clean and tidy state before it has even more seedlings and new plants growing in it. Tidy greenhouse benches, then give them a clean. Throw away all dead leaves and petals as they can harbour diseases. Brush all dirt from surfaces, then wipe with mild disinfectant if traces of dirt are left. To keep the atmosphere dry, avoid leaving any damp patches by wiping them dry with a cloth afterwards.

JOBS TO DO THIS MONTH

- Keep control of early greenfly. Look out for the sticky, translucent clusters of greenfly feeding on your plants' newest buds. Rub them off straight away and follow up with a wash of insecticidal soap or plant oil to control the young nymphs that are easily missed.
- Ventilate in good weather during the day to save plants from exposure to extremes of temperatures.
- Harden off hardy seedlings sown last year using a cold frame.
- Feed and water actively growing plants.
- Sow lettuces into small pots or modules for planting out in a few weeks' time.

Fresh from the Garden

Although the warmer weather is good, the weeds also enjoy it. As well as tackling the weeds, it's time to dig over the veg beds and mulch the areas where you'll be growing crops, with homemade compost. It won't be long before we can harvest first baby leaves of the salads that have been sown indoors or the fresh vibrant colours of micro greens such as alfalfa, radish and sweetcorn.

WHAT TO SOW

Indoors:

- Cucumbers
- Lettuce

Outdoors:

- Broad beans
- Brussels sprouts
- Peas

WHAT TO PLANT

- Strawberries
- Blueberries

WHAT TO HARVEST

- Cauliflowers
- Cabbages
- Forced rhubarb
- Sprouting broccoli
- Swiss chard

On the Veg Patch

KEY TASKS

- Plant onion sets.
- Pot on tomatoes.
- Sow beetroot.

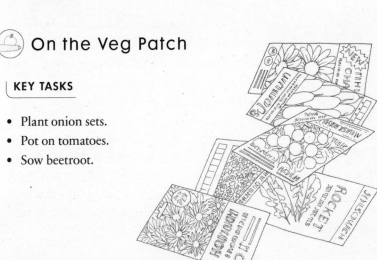

CROP OF THE MONTH: FORCED RHUBARB

Did you know? – The pinker the stem, the higher the contents of micronutrients and antioxidants.

Nutrients – Rhubarb is a good source of vitamin C, vitamin K, magnesium and fibre.

Storing – Place freshly cut stems inside a plastic bag to prevent them drying out and store in the fridge for a few days. For longer storage, wash and cut the stems into chunks and seal in a bag. These can be frozen for up to a year.

Good cultivars – 'Victoria' for heavy yield; 'Timperley Early' for early cropping.

Plant – November–March

Harvest – February–July

HERB OF THE MONTH: PARSLEY

A versatile herb that is good for bringing out the flavour of other foods and herbs. Add just before the end of cooking. Curly leaved varieties are more often used as garnish. Flat-leaved parsley has a stronger flavour, so is favoured by many chefs.

Sow indoors with heat from late winter or outside in late spring. Very slow to germinate, but a warm environment and damp soil should speed this up considerably. Plants are ready to harvest just a few weeks after sprouting. Prefers a rich, damp soil in partial shade, so dig in organic matter before sowing. Very good in pots, both inside and out.

GOT FIVE MINUTES?

Hoe between veg rows

Tiny weed seedlings develop fast to compete with the emerging rows of vegetable seedlings. On a dry day, use a hoe to cut through even the tiniest weed seedlings that pop up between the rows. Done regularly, this job will take no time and your vegetable patch will be clear of invasive weeds. If you keep the edge of the hoe sharp with a file, the job will be even quicker.

GOT HALF AN HOUR?

Prep your strawberry bed

Get your strawberry bed ready for this year's harvest. Weeds will have grown under the old foliage by now and this crop does so much better with less competition for moisture and nutrients. It's a good idea to mulch your beds with straw to keep the berries clean, dry and easier to spot for harvesting.

- Go through each plant and trim off any dead or old leaves. Pull away winter debris such as moss and old mulch from the crowns.
- Gently hand fork through the bed to remove weeds among the plants without disturbing the shallow-rooted strawberry plants. Make sure you lift the roots of perennial weeds.
- Give your crop a big nutrient boost with a top dressing of organic fertiliser scattered on the soil around each plant.

GOT A MORNING?

Plant asparagus

Although asparagus is considered a delicacy, it's easy to grow. However, you must be patient. After you've planted your roots, wait at least a couple of years before harvesting, and even then harvest lightly to begin with. This gives them time to establish, leading to large, juicy shoots when you come to cut the crop. After that they will produce a crop every spring for at least 15 years. Harvest with a sharp knife when the spears are 15cm high, cutting just under the soil. Very little maintenance is needed:

protect against slugs and snails, keep weed free and ensure there is good drainage. Only cut the spears for two months each year, then allow the ferny fronds to grow over summer, tying back as needed. When they go yellow in autumn, cut them back to just below the soil.

1 Site your asparagus bed on free-draining soil. If you don't have that, create a raised bed. Then dig a 20cm-deep trench and fork well-rotted manure into the base.

2 Create a 10–15cm-high ridge of soil to sit the crowns on, to protect them from waterlogging – I sometimes add a little grit. Firm the ridge in place so that it's stable.

3 Dangle the crowns over the ridge so that they sit either side of it, 30cm apart. Cover the roots with soil but leave the tips of the buds just sticking out.

4 Water carefully, ensuring you avoid disturbing the position of the crowns. Top-dress with fertiliser in spring and mulch with organic matter every autumn.

TOMATO FOCUS

Tomato types explained

- **Cherry** The smallest and often the sweetest of tomatoes. They are ideal for adding whole or halved to salads, pasta dishes and kids' lunch boxes or for roasting whole.

- **Cherry plum** Also called baby plum, these are small and sweet like cherry tomatoes but with an elongated shape.

- **Salad** The classic medium-sized tomato. These are good chopped into salads or for grilling, baking or frying.

- **Plum** Similar in size to salad tomatoes but with an oval shape. They are fleshy, with fewer seeds, making them ideal for turning into sauces and soups, as you don't have to overcook them while waiting for the juice to evaporate.

- **Beefsteak** The biggest tomatoes. Beefsteak tomatoes have a meaty texture and are good for grilling, stuffing and slicing for sandwiches.

Top tips for growing your best-tasting toms

- Choose a recommended variety to ensure you get a great flavour.
- Light is more important than heat for developing flavour, so ensure your tomatoes are in the brightest spot possible. If growing in a greenhouse, adjust the vents to try to keep the temperature between 21°C and 24°C. Temperatures over 27°C can cause the plants to suffer and mean you'll have to water a lot, which can dilute the flavour.
- Sow your tomatoes early or get a head start with plug plants, especially if you're growing under glass. You'll get longer cropping.
- Fast-growing cherry and plum varieties ripen earlier than large varieties, and so make it easier to get a good flavour in dull summers. If you're in the north of England or Scotland, smaller varieties are a safer bet.
- Look out for pests and tackle them as early as possible.
- Feed fortnightly with a potassium-rich fertiliser when flowers appear but don't overdo it. Overfeeding results in growth problems.

Varieties with an AGM

- **'Favorita'** This is a cherry tomato with a superb sweet flavour. Early, prolific crops on long trusses and with good resistance to multiple diseases.
- **'Suncherry Premium'** A vigorous cherry variety that produces very long trusses and huge crops of tiny, uniform deep red fruit that keep well on the plant. A good sweet/acid balance and depth of flavour. Best grown in the greenhouse.
- **'Sungold'** This cherry tomato has a deliciously sweet flavour due to high levels of beta-carotene and sugar. Good yield, early to mature and good disease resistance. Prone to split skins when over-ripe, but it's so tasty not many get that far!
- **'Olivade'** While this juicy plum tomato eats well fresh, it comes into its own when cooked. Being French it's better suited to the north Europe climate than Italian plums. Reliable, with good disease resistance.

> - **'Brandy Boy'** An improvement on the heritage variety, the beef-steak tomato, 'Brandywine', has increased vigour and better fruit set, improved disease resistance and earlier yields. Reliable with a rich flavour and a meaty texture.
>
> *'Once you've enjoyed the burst of flavour from a just-picked, home-grown tomato, still warm from the sun, it spoils your experience of shop-bought ones forever.'* – **Sally Nex**

OTHER JOBS

Harvest the last of the winter veg

At this time of year, winter vegetables left in the garden will run to seed, ruining any chance you had of eating them. Leave some for seed if you are a seed saver, but harvest the rest now to enjoy the last of these hearty veg. Winter crops get a good dose of autumn mulch, so the ground where they've been growing will already have a lovely fine tilth ready for sowing. Rake in some fertiliser before preparing the seedbed for spring sowings of salad crops or planting onion sets.

Start onion sets

Plant onion sets straight into prepared ground now. Sets are supplied as small, dormant, immature onions, which will root in and grow away fast. An alternative way to grow onions is to have sown seed under glass earlier in the year and grown on the young plants before planting them out into the veg bed. Rake the soil level, then space the sets 10cm apart in the row, with rows 30cm apart to maximise root growth. Using a trowel, plant the bulbs by sitting them pointed end up and firming in with just the top above the surface. Never press the bulb in hard as you risk damaging the base. Keep your eye on them as they establish and replant if birds pull them out.

Get ready to plant out

Move young plants that are ready for planting out from the warm green-house to acclimatise them to life outside. If you plant them straight out from a warm environment they will be susceptible to leaf scorch from the cold and the drying winds. Put them in cold frames or a sheltered spot outside, where they will be protected but will gradually harden to the lower temperatures. Open the lids during the day to introduce some air movement. If you don't have a frame, they can still go outside but it is wise to move them back in or cover them at night for the first few days. They'll establish well after this process, which should take one to two weeks.

TRY SOMETHING NEW...

A different variety of coriander

Coriander 'Leisure' is an intensely flavoured herb, and this variety has been specially bred to produce lots of big leaves and be extra slow to bolt. **How to:** Coriander grows well in rich soil that's previously been mulched. Rake the growing area to a fine tilth and broadcast the seed. Lightly cover by raking over soil. Water in well, and cover with fleece until germination takes place. Harvest when the plants produce six to eight stems.

 # Rekha Mistry's Recipe of the Month

RED LENTIL AND LEEK SOUP

Serves 4

You'll need

100g split red lentils
500ml vegetable stock
100g chopped leeks
100g chopped tomatoes (fresh or tinned)
1 small onion, diced
1 green chilli, finely diced
1 tsp grated ginger
2½ tsp sambar spice powder (available online and from Asian
supermarkets)

½ tsp Kashmiri chilli powder
½ tsp coriander powder
salt to taste
juice of half a lime
For sambar
2 tbsp sunflower oil
2–3 dried whole chillies
1 tsp black mustard seeds
4–5 curry leaves

Method

1 Wash the lentils, then place in a pan with the stock and cook until tender – about 20 minutes.
2 Add the leeks, tomatoes, onion, chilli, ginger and all the dried spices, and cook down for a further 20 minutes (if you want a smooth soup, give it a whizz in a blender).
3 In a separate small pan, heat the oil, add the dried whole chillies, mustard seeds and curry leaves. As soon as everything starts to sizzle and splutter, carefully pour the hot oil and spices mix into the soup, and stir well.
4 Add salt to taste, squeeze over the lime juice, then serve hot.

 ## Wildlife Notes

March brings the deep, low buzz of queen bumblebees flying in low zigzags as they search for an old mouse hole to make a nest. If you have crocuses, then look out for bees feeding on their pollen and nectar. Sometimes bees fall asleep in the blooms overnight. Look out for big clumps of frogspawn at pond edges; and peacock and small tortoiseshell butterflies sunning themselves on bright days. Birds are singing loudly from the rooftops – listen out for the melodious song thrush and blackbird, which join the robin and great tit, who have been in voice for a few weeks. Keep bird feeders topped up, so the birds have all they need to raise young, hopefully in your garden.

WILDLIFE WATCH

Look out for bumblebees lying on the ground. These bees are valuable early pollinators. If you see one, move it on to a flower so it can replenish its energy, or you can mix a sugar solution by mixing equal parts warm water and sugar. Place it near the bee's head in a bottle cap or something similar, and it should stick out its proboscis to drink, energise and warm up.

LOOK OUT FOR...

- Newts eating frogspawn in your pond; they'll soon start to breed. Shine a torch into your pond at night – you might see males using courtship rituals like 'tail fanning' to send pheromones over to waiting females.
- Brimstone butterflies, which are some of the earliest butterflies to fly in spring. See if you can distinguish the lemon-yellow male from the greenish-white female.
- Hairy-footed flower bees. They look like bumblebees but zip around like hoverflies. The female is black with orange legs and the male is pale ginger. They feed on lungwort, primrose, comfrey and dead-nettles.
- The UK's smallest vole, at just 8–12cm in length, the bank vole is also the most common and therefore the most likely species to visit gardens.
- Slow worms. On sunny days you might spot them basking beneath a large stone to warm up.
- Birds with twigs in their beaks. Nest building is well under way for many species. Leave out some soft nest material, such as dog hair, for them to take.
- Hoverflies – a few species, such as the drone fly *Eristalis tenax*, are on the wing now.

🔭 Spotter's Guide to Amphibians and Reptiles

Though little else is moving, this is the most active month for amphibians, which start their return to the water where they will breed. They spend only a few months in ponds, just to mate and lay their eggs. This often happens in a frenzy, as numerous males struggle to grip a smaller number of females. Once they have mated, the adults soon move away; by late summer, even the tiny young are no longer aquatic tadpoles and they, too, migrate to land. The name amphibian means both kinds (*amphi*) of life (*bios*) – terrestrial and aquatic. Reptiles are not tied to water, though grass snakes like to hunt in it for fish and occasionally frogs. The fermenting warmth of compost bins and manure heaps are more attractive to reptiles, which use the heat to lay and incubate their eggs and young.

LOOK OUT FOR...

- **Common frog** – This frog has a smooth-skinned, greenish, brownish or yellowish body of 6–9cm, often marked, marbled or streaked. It has long legs for hopping, and large black eyes surrounded by gold, with dark, triangular markings behind each eye. Lays over 1,000 eggs in a gelatinous spawn mass.
- **Common toad** – This toad has an 8–15cm, large, stout, warty, brown or olive-green body. Fatter and larger built than the frog, its short legs are used for walking. It has small eyes and lays 600–4,000 eggs in gelatinous spawn ribbons. Long-lived, with 40 years recorded.
- **Slow worm** – The slim, pale, metallic, greyish-brown body of this legless lizard reaches 30–45cm. Dark-streaked down the sides, especially females and young. Often shelters in groups in warm, dry places. It releases its tail to escape predators – the tail continues to writhe long after.
- **Smooth (or common) newt** – This newt's slim, dark, mottled, olive-green, brown and black body reaches 7–10cm, including tail. Males have a brightly patterned orange belly with black spots, and a tall,

wavy crest down the neck, back and tail. Females look more demure. Lays eggs singly on aquatic leaves.

- **Grass snake** – This snake's olive-green body, 70–150cm, has dark spots along the sides. Its head has a yellow and black collar with large, black side marks behind. The underside is a black and yellow-green harlequin pattern. Visits ponds to hunt fish, frogs and newts. Swims, leaving a sinuous wake.

 Troubleshooting Guide

DID YOU KNOW?

Garden soils comprise three types of particles – sand, silt and clay (here in descending size order). Average garden soil is composed of around 50 per cent rock-based minerals and organic matter, with the other 50 per cent being water and air. Clay soils are so tightly packed that their capacity to hold air and water is well below the typical 50 per cent.

GROWING IN CLAY SOIL

If you're hankering after plants that need sharper drainage than clay can provide, there are several solutions. Digging plenty of organic matter, ideally well-rotted manure, deeply into the soil will open it up, improving drainage and helping it to warm up earlier in the season. Adding grit, too, will go some way to improving drainage, but it's no silver bullet. You can also apply organic mulch to the soil surface, which will be pulled down into the substrate by soil organisms over time, opening up the texture and improving drainage. Or you can apply lime if the soil is acidic. This improves the clay by a process known as flocculation, which essentially means the finest particles clump together, forming larger air and water pores between them, thus improving drainage, workability, air movement and warmth. If you're prepared to do a little work to improve the soil, then you'll be able to grow a wide range of beautiful clay-loving plants.

USING MYCORRHIZAL FUNGI

These soil organisms, whose name comes from the Ancient Greek words for fungus and root, bind onto plant roots, increasing the root area and absorbing water and mineral nutrients that can be used by the plants above. Mycorrhizal fungi are specialists, able to take up nutrients in soil that plant roots cannot access. In return, plants supply them with carbohydrates that they've produced by photosynthesis. Garden soils may lack mycorrhizal fungi as a result of cultivation and the use of fertilisers, but we can add them when we plant. With long-term plantings, such as trees and shrubs, it's a good idea to dust the rootballs with mycorrhizal fungi (look for Rootgrow) just before covering them with soil. Our understanding of these ancient organisms is limited, but research has shown that plants acquire better resistance to drought, diseases and pollution through this relationship, which even allows them to transmit information to each other.

TOP TIP

Apply mycorrhizal fungi as close to the roots as possible when planting, as they need to be attached to roots to survive.

PATCHY LAWN

As with all gardening, the best lawn care is guided by science. Botanically, lawn grasses are like all other plants in that they need food to survive and grow, and will only tolerate limited competition. But unlike some other plants, they respond well to being cut frequently by producing more sideshoots (known as tillering). This means that mowing a lawn makes it fuller. You should start mowing early in the season then do it often (but avoid mowing very damp ground). Mowing also reduces weed competition – lawn grasses cope with it better than most weeds because their growing point is too low for the mower blades to damage.

Specially formulated spring lawn feeds are high in nitrogen, a crucial component of chlorophyll, the photosynthesis chemical, which is stored in verdant grass blades. So by giving the lawn a feed at the end of this month you can compensate for all the chlorophyll lost in the clippings you'll remove in the grass box this summer.

TOP TIP

Want to mow dewy grass? Knock excess moisture off the blades by drawing a broom over the top of it.

DUCKWEED ON PONDS

Duckweed can be a big problem in garden ponds, as it spreads so rapidly in warm weather. The best way to control it is to skim it from the surface with a net. This shouldn't upset resident frogs, and if you leave the weed at the side of the pond for a few hours, any small creatures that have been scooped up can crawl back in. The weed you remove can then be composted. To reduce any recurrence, there are a few things to try. In a large pond you could add grass carp, which are herbivorous fish that eat duckweed. You could also increase the level of shade falling on the water as this will hinder duckweed growth – try planting moisture-loving shrubs such as ornamental willow.

FRUIT FOR A NORTH-FACING FENCE

The further north or west you are, the harder this will be, with fewer choices that succeed easily. Some trained fruit can do well on a north wall as long as the site is not shaded – if open to the sky in summer, it will get considerable light especially at early and late times of day. Red and white currants are excellent, easily trainable and very productive. Gooseberries likewise, as long as there is also good airflow. Although not espaliered as such, blackberries can be similarly trained and are worth the effort, as can the tayberry, which can do better on a shady rather than sunny site. Pears, which crop later than most other fruits, were traditionally trained in such positions to extend their season and, where given more height than an average fence, so were Morello cherries.

DEALING WITH SUCKERS

Suckers grow from root systems and, while they can be useful for propagation, more often they're a nuisance. For a plant, producing suckers is a good way to reproduce, by making a genetically identical copy of itself. 'Adventitious' buds on the roots grow into shoots, usually next to but sometimes quite a distance from the original plant, annoyingly often in a lawn, path or border.

Certain plants, particularly those with shallow root systems, sucker more readily, such as bay, cherry, hazel, lilac, plum, poplar, privet, robinia and stag's-horn sumach. Grafted plants, for example roses, are also prone to producing suckers, which are easy to spot as they have different leaves from the main plant. Some growing conditions, such as a rocky subsoil or a high water-table, can make suckering more likely, by encouraging the roots to come to the surface. Any gardening that damages roots may cause it too.

To control suckers, scrape away the soil and try to pull them off, so you remove the adventitious buds from which they grow. For larger areas of shallow roots with suckers, after removing them, you can try to discourage them by raising the soil level.

APRIL

April has fickleness built into it. The only thing that is certain is that it will be unpredictable, variable and contrary – and that is its charm. Certainly, no other month changes so much from the first day to the last. It is also a busy month, the days are stretching out, the soil is warming up, and the garden calls like a siren. It is a month spilling over with energy and hope. Across the land, vegetable plots will get their first real attention. It is also the ideal time to start a herb garden, and to renew and replant herbs that have suffered over winter as well as get planting for summer, everything from sweet peas to summer-flowering bulbs.

Weather Watch

It may come as a surprise that the month traditionally associated with showers is actually one of the driest of the year. The transition of weather patterns from winter to summer mode starts to slow and weaken the jet stream driving the Atlantic storms that bring us most of our winter rain. With fewer, weaker depressions reaching us, most of our April rainfall comes from scattered showers that deliver much smaller amounts of rain overall. Showers are formed when cold air is warmed from below, generating rising bubbles of air that cool and condense into shower clouds. In winter, the seas provide that 'bottom heat', but in April the strengthening sun warms the surface of the land sufficiently to produce showers and we start to notice them more, hence 'April showers'.

WEATHER FACTS

- Highest average rainfall: northern Scotland, 16.1 days
- Lowest average rainfall: Yorkshire, 9 days
- Average high temperature: 11.4°C
- Average low temperature: 3.4°C
- Lowest temperature: East Anglia, 2.7°C
- Most hours of sun: the south east of England, 180.3

DID YOU KNOW?

Some regions of the UK such as the Western Highlands in Scotland receive such high levels of annual rainfall, up to 300cm, that wooded areas in these regions are classified as rainforest.

WEATHER ALERT

With soil conditions improving through the month, it's worth preparing for a dry spring by conserving precious moisture left over from winter. Organic matter is the key, so incorporate as much as possible while soil is still moist. A surface mulch will reduce loss from evaporation.

 Star Plants

10 OF THE BEST

1. Muscari

These small, spring-flowering bulbs provide glorious colour. Also known as grape hyacinths because their flowers look like tiny bunches of grapes, they are perfect for containers. They are very reliable and easy to grow, but spread easily. Flowers April to May.

2. Snakeshead fritillaries

These nodding chequered flowers with viper's tongue leaves are eye-catching spring flowers. For the best effect, plant lots of them in long grass or on the margins of shrubberies. If you only have a small garden, then plant a big potful that you can see from the kitchen. Flowers April to May.

3. Epimediums

Epimediums are an appealing and increasingly popular group of low, spring-flowering hardy perennials. Essential ingredients of any shady garden, they mingle well with other early flowers to form an enchanting spring tapestry. Their intriguing spidery flowers may look delicate, but these plants are tough – some will even thrive in dry shade. Try *Epimedium x perralchicum* 'Fröhnleiten', which are good for dry shade, or the pink and yellow 'Fire Dragon'. Flowers April to May.

4. Tulips

There are 15 groups of tulips. Triumph tulips are single tulips that flower from about late April, after the Single Earlies and before the Single Lates. The stems are about 40–60cm tall, supporting the flowers well, and the flowers themselves have that familiar, classical, elegantly egg-shaped form. There are over 600 of them in an extraordinary range of brilliant, rich and pastel shades. Try 'Prinses Irene', which has orange flowers streaked in smoky purple, or 'Jan Reus' with rich, deep red flowers. Flowers late April to May.

5. Erythroniums

This glorious inhabitant of the woodland fringes is overburdened with common names – dog's tooth violet, for one, because of the shape of the bulb. This flower will put a sparkle into anyone's day. It needs shady woodland, although not too close to trees that may vacuum up all the summer moisture. Plant in drifts where they can spread undisturbed. Try the lovely 'Citronella' which has bright lemon flowers from March to April.

6. Candytuft

Candytuft, *Iberis sempervirens*, is one of those plants that was in everybody's grandmother's garden. It was something (a bit like aubretia) that just turned up without anybody actually going to the trouble of propagating or buying it. Maybe it swept into gardens under cover of darkness. Try 'Snowflake' which is evergreen and makes a good rockery plant. Flowers April to May.

7. Trilliums

Trilliums are spring-flowering woodland plants, native to North America and Asia. Some have bright flowers and variegated foliage, others have fresh green leaves and white flowers. They thrive in sheltered, shady gardens in moist but well-drained soil. Try *Trillium erectum*, which has three glossy leaves and one perfect 'clarety' flower, comprising three sepals and three petals. Needs a good, rich soil – some leafmould in autumn would not go amiss. Plant in dappled shade and keep an eye out for slugs and snails. Flowers in April.

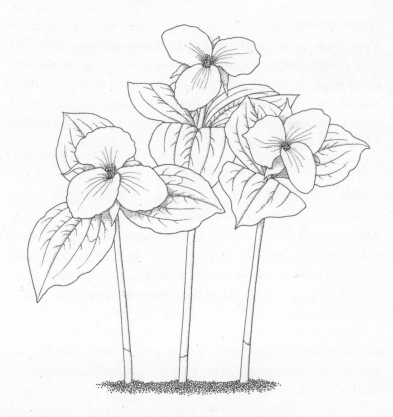

8. *Kerria japonica* 'Pleniflora'

This is a very easy shrub to grow, and perfect for those corners that just need something straightforward and manageable. Its common name is Bachelor's buttons but 'Pleniflora' means 'lots of petals' – it has double yellow flowers.

It will thrive in most places apart from deep shade. Flowers March to May.

9. Bleeding heart

You may remember this plant fondly as dicentra; it is now known as *Lamprocapnos spectabilis*, and it's a treasure for a slightly shaded border. It's a Chinese plant bearing pink-red, heart-shaped flowers with white tips, which hang from arching flower stems in late spring to early summer. Flowers April to June.

10. Siberian bugloss

If you're looking for something more unusual for the spring garden, try *Brunnera macrophylla* 'Jack Frost'. This attractive plant is covered with lots of small, bright blue flowers which look a bit like forget-me-nots. These are complemented by the beautiful heart-shaped foliage which is a pale silver-grey with pronounced dark green veins. Flowers April to May.

NOW'S THE TIME TO...

Plant up a hanging basket

A satisfying way to celebrate this feverish time of year is by teaming seasonal favourites in good-looking pots or baskets and displaying them where you'll see them often. Garden centres are awash with tempting spring plants. Plant up your hanging baskets this month and let them establish under protection before being hung outside once the risk of frosts is over. Part fill the basket with potting compost, then position a single focal plant, such as a geranium, in the middle. Add a selection of trailing plants around the edge and some free-flowering fillers in the gaps. Water the basket well, then hang it up in a light but frost-free spot to establish.

PLANTS FOR HANGING BASKETS

- Petunias
- Violas
- Creeping Jenny
- Calibrachoa
- Lobelia

WHY NOT TRY...

Edible flowers

- **Primrose** – crystallise and use as a decoration on cakes
- **Viola** – crystallise and use as a decoration or scatter petals on salads
- **Courgette** – remove stamens, stuff with cream cheese and fry
- **Cornflower** – clove-like flavour and attractive as a salad garnish
- **Lavender** – use sparingly in cakes, biscuits and lemonade

Jobs to Do this Month

Spring is in full swing – it's a great time to be outdoors in the garden and there are no shortage of jobs to be done, from the small tasks such as checking for pests to the bigger ones like planting a hedge or reviving the lawn. Big or small, these jobs are a satisfying part of the gardening calendar, at the beginning of a new season that is full of hope.

KEY TASKS

- Plant dahlia tubers outside.
- Plant autumn-sown seedlings outdoors.
- Plant summer-flowering bulbs, corms and tubers.
- Prune back lavender and penstemon to fresh new shoots.
- Keep potting on seedlings to avoid a check in their growth.
- Feed the lawn with a high-nitrogen feed to promote strong growth.
- Trim winter-flowering heathers.
- Turn the compost heap.

- Keep tying in climbers as they grow.
- Order nematodes for specific pests.
- Water containers.

GOT FIVE MINUTES?

Move seedlings

Look for self-sown annual seedlings springing up. These can be weeded out or moved now. They should establish well and flower later in the year. Water seedlings thoroughly after replanting.

GOT AN HOUR?

Set up plant supports

Flowers with tall stems or annual and herbaceous climbers need supports put in place, ready to train in early growth. Tying in shoots or propping them up later risks damaging the plant and never looks as natural. Clean up any structures or poles you've stored over winter, then use them (or new prunings) to make wigwams or arches. Make sure the frame supports the stems low down but keeps the flowers freestanding. For plants in pots, firm the frame securely down into the pot.

GOT A DAY?

Plant a hedge

Hedges add structure, define border edges and create new areas. For a formal-shaped hedge, trim the tops of plants like a *Buxus* to the level of last year's growth soon after planting. Leave informal hedges like lavender or fuchsia to make their natural form and prune as you would a single specimen.

1 Choose containerised hedge plants for planting now, such as yew or holly. Box is a traditional option but may suffer from pest and disease problems.
2 Set out the plants according to your chosen spacing. Base the decision on plant vigour, budget and desired effect. Plant closely for a straight, formal hedge line.

3 Keep the rootball surface at ground level as you plant the hedge and firm either side of each plant with your hands or feet as you work down the row. Avoid stem rot by stopping the soil from building up too close to the woody stems.

4 Soak the plants along the row before applying mulch on the ground either side of the new hedge.

AND A FEW OTHER JOBS...

Plant out sweet peas

Knock sweet peas out of their pots and plant them into a sunny border or pot with rich, moisture-retentive soil or compost. They will need a structure of pea sticks or netting they can scramble up to flower through the summer. Keep them well watered as they establish, and pinch out the tops if they are getting leggy. Sweet peas are hardy enough to cope with some frost but need protection from slugs and snails.

Plant summer bulbs and tubers

Plant exotic bulbs and tubers now, for their dramatic flowers later in the summer. Dahlias, cannas, agapanthus and gladioli can all be planted now. You may still find dormant bulbs this month but also look for them growing in pots at the garden centre. These will mostly need sun and a free-draining soil. Mulch with grit to keep bulbs from rotting and protect the new shoots from slug damage.

Check for aphids

Early infestations of aphids can quickly get out of control if they go unnoticed. Check plants regularly and, if aphids appear, squash by hand while numbers are still small. Encourage natural predators like hoverflies, ladybirds, house sparrows and blue tits into the garden to feast on the tiny sap-sucking insects.

Add pond plants

April is an ideal time to add new flowering plants to the pond. You can also divide aquatic plants this month, apart from waterlilies and water irises which are best divided in May and July respectively. Propagating by division is a simple and cost-effective way to increase plant numbers in your garden pond.

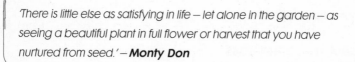

There is little else as satisfying in life – let alone in the garden – as seeing a beautiful plant in full flower or harvest that you have nurtured from seed.' – **Monty Don**

SOW SEEDS

Give seeds the space they need with the right-sized container and keep moving the seedlings on so they never get overcrowded or slow down.

1 **Into trays** – Sow tiny seeds, like delphinium and campanula into trays. Place in your palm and tap gently to scatter evenly over compost. Cover thinly with sieved compost. Once true leaves appear, prick out.
2 **Into modules** – Medium-sized seeds such as French beans, zinnias and calendula are best grown in modules. Simply poke a hole in each and drop the seed in, pinching the compost over to cover.
3 **Into small pots** – Large flower seeds like cobaea, ipomoea and cosmos can be grown in small pots to avoid being repotted often; just sow as you would into a module. Pots made out of newspaper can be planted directly into the ground.
4 **Into large pots** – Larger pots suit fast-growing plants like sweet peas or sunflowers. Make two holes, 1.5cm deep, in each pot and sow one seed per hole, pinching soil over to cover. Plant outside once roots grow out of the base of the pot.

WATER WELL

Feel the compost or lift and feel the weight of your pots and baskets to check whether they need watering. Plants putting on leafy growth require more water. Warm or windy conditions will increase the need for watering too. Don't rely on the rain to do the job as leaves act like an umbrella and the rain ends up on the ground around the pot.

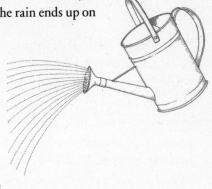

Soak the compost by holding the watering can close to the compost surface as you pour. If pots are very dry, go back and water them again to give the water a chance to soak right through, rather than run off down the sides of the pots.

CONTROL SLUGS AND SNAILS

Mix a solution of nematodes with water in a watering can and water the soil around plants to control slugs and snails. They are prey to these microscopic worms that are active when ground temperatures reach 7°C. Use plenty of water; it helps nematodes to move around the soil. The effect lasts for around six weeks and it's good to repeat it, but used now it's very effective as it controls slugs during egg laying.

DON'T FORGET TO...

- Check fruit cages and netting for holes that will allow pests in.
- Give the lawn a boost with a high-nitrogen spring feed.
- Plant out autumn-sown hardy annuals.
- Feed shrubs and hedges with bonemeal.
- Mow the lawn every two weeks.
- Water pots regularly.

LAST CHANCE TO...

- Cut back hardy shrubs such as buddleia. Woody shrubs that require a hard annual cutback, such as buddleia, are usually dealt with in March, but you can get away with it now, too.

Time to Prune

With winter's dark days long since passed, April sees the garden coming back to life, with new buds and blooms bursting out everywhere. The air and soil have warmed up, letting plants know that now is the moment to go for it. The garden is also teeming with other life, as birds and insects set about their spring duties. A wide range of plant groups need pruning this month. Tender and subshrubs are now safe to tackle as the worst of the weather has subsided, but grasses, hydrangeas and other hardy shrubs can also be addressed. It's also an ideal time to tidy up low groundcover species such as winter heathers and *Hypericum calycinum*.

PLANTS TO PRUNE

- Hardy fuchsia
- Abutilon
- Sedum
- Ferns
- Sedum
- Ornamental grasses
- Camellia

HOW TO PRUNE... MOPHEAD HYDRANGEA

These ever-popular shrubs produce the best blooms at the tops of strong stems from July onwards. Left unpruned, they can take up lots of space for very little in the way of flower, developing a mass of woody unproductive stems. This is the month when their big buds start into growth. To promote a good show of blooms next year, take out the thinnest and oldest flowered stems now, as this will encourage new, strong shoots from the base.

1 Examine the shrub, looking for any broken or snapped branches and cut these out first, pruning to just above a healthy pair of buds.

2 Use secateurs to cut out one or two of the oldest, thickest stems to just above ground level and cut out any shoots thinner than a pencil.

3 Remove the old flowers that have protected the plant over winter, cutting back to the second or third pair of buds below the flower.

In the Greenhouse

It's a busy month in the greenhouse. There's plenty to do – keeping watch for pests such as vine weevil, sowing new plants from herbs to cucumbers and potting up plugs for summer displays. Keep plants healthy by shading new plants from direct sun and ventilating the greenhouse on warm days.

JOBS TO DO THIS MONTH

- Pull the shade blinds down when the sun shines to reduce the chance of soaring temperatures and direct sunlight scorching young seedlings. If the seed leaves are damaged the newly emerged seedlings will die at this vulnerable stage of their development. It's worth keeping all plants at the propagation stage out of direct sun to protect the new growth. If you don't have blinds, shade netting or fleece pinned on the inside of the windows work just as well.

- Sow seed of leafy herbs like basil, coriander and dill. They're expensive bought in pots as part of your weekly shop, but they are really easy to grow. You'll need to sow more every couple of weeks so that you always have plants at just the right stage in the kitchen for cutting.

- Pick off adult vine weevils if you see them. Each one can lay thousands of eggs now and the new grubs eating plant roots are the most damaging. Discard compost and repot into fresh if you have an infestation. A combination of hand picking, using a nematode drench over compost and traps for the adults should keep on top of them.

- Young plants, such as seedlings or bedding potted on from plugs, cannot tolerate strong sunshine. Sunny days are devastating in the greenhouse if they're not protected from sun rays, as their leaves are usually delicate and water loss can be rapid. Shield trays and pots with cardboard propped up to provide shade, or rig up shade netting. Check compost moisture daily.

- The risk of frost will soon be over in most parts of the country, and you'll be able to start planting tender plants outside. But they'll suffer a setback if you put them straight outside – instead you need to harden them off by putting them out during the day then covering them at night. By introducing them gradually to lower temperatures and more air movement, you'll toughen them up for the season ahead.

Fresh from the Garden

April is the most thrilling month in a gardener's diary. Everything, including the weeds, seems to come alive the minute the warming sun rays hit the ground. There's so much to do: beds to prep, and seeds to sow – both in the greenhouse and direct outdoors, where you can sow cool season crops like peas, lettuce, radishes and spring onions. Then there's seedlings to pot on and first early seed potatoes to plant out.

This month's to-do list for growing your own is long but satisfying. It includes turning the compost heaps and leafmould. Broad beans that were sowed in autumn now need staking and a routine aphid pest check. It's time to prepare a compost-filled trench for runner beans and direct-sow peas. Later in the month it's time to start hardening off crops grown in the greenhouse such as courgette, cucumber and tomato. They can be moved to a cold frame before planting out mid-to-late May.

Another job is to remove all winter veg, like cabbages and kale, as the warmth has started to make them go to flower. Remember it's a good idea to change planting areas for vegetables – because this process of crop rotation is a natural way to avoid a build-up of pests and diseases.

You could try sowing calendula and tagetes seeds to act as natural pest repellents. Greenfly seem to hate their scent, so they are good grown with tomatoes – and they attract pollinating insects.

WHAT TO SOW

- Brussels sprouts
- Kale
- Peas
- Spinach
- Cabbages

WHAT TO PLANT

- Potatoes, second early and main crop varieties
- Artichokes
- Asparagus
- Onion sets
- Strawberries

WHAT TO HARVEST

- Asparagus
- Beetroot
- Radishes
- Rhubarb
- Salad leaves

 ## On the Veg Patch

KEY TASKS

- Keep frost off early blossom.
- Sow courgettes.
- Mulch fruit bushes.
- Harvest rhubarb.

CROP OF THE MONTH: PURPLE SPROUTING BROCCOLI

Did you know? – The leaves are also edible and full of vitamins. Try steaming, stir-frying or microwaving rather than boiling to get the most nutrients from your broccoli.

Nutrients – Broccoli is a good source of vitamins C, K and B6. It also contains calcium and dietary fibre.

Storing – Cut spears can be stored in a bag inside the fridge for a few days. If freezing them, blanch first.

Good cultivars – Broccoli 'Purple Sprouting Early' for a longer spring harvest; Broccoli 'Red Arrow' for extra-tender spears.

Sow – March–June

Harvest – February–May

HERB OF THE MONTH: CHERVIL

The leaves are much loved in French cooking and are best added just before serving to preserve their flavour. They are delicious with chicken or fish, and in soups. Look for any seed sold as chervil. There are no specific named varieties.

Grow it in pots or multi-cell trays in a cold greenhouse from late winter, or sow directly outside from early spring. It dislikes being moved. It needs some shade to stop it from flowering too readily, so is good inter-planted with taller plants and prefers a light, moist soil. It can be grown outside in pots as a cut-and-come-again herb, but gets too weak and leggy on the kitchen windowsill. Pick out any flowering growth to ensure it keeps producing leaves for harvesting. A summer sowing will provide fresh leaves throughout the winter but cover plants with a cloche in autumn for protection. It will germinate quickly in warm weather and plants can be ready to harvest six weeks after sowing.

GOT TEN MINUTES?

Top dress containerised fruit

Scrape away and discard the top layer of compost first. Replace this with new compost mixed with a dose of fertiliser, then water so the roots can absorb the nutrients.

GOT HALF AN HOUR?

Sow carrots

Now is the perfect time to sow, since carrots need to grow rapidly and without check if they are to be sweet and tender. The task is easy, as they need nothing more than watering in dry spells before they are pulled and enjoyed a couple of months after sowing. Sow too thickly and you'll have to thin out the young carrots, which is a mistake since the aroma released will encourage carrot fly whose larvae burrow into the roots. Sow thinly and with even moisture and they will seldom encounter any problems.

- Sow in fine or sieved soil to avoid the roots forking. Firm and level the soil and make a shallow drill by pressing a cane into it.
- Sprinkle seeds into the drill so that they land between 1 and 2cm apart, before raking back the soil. Sowing thinly will avoid the need to thin and attract carrot fly.
- Give the seeds a head start by protecting them from the worst of the spring weather with a homemade tunnel. It's not essential, but is beneficial in exposed areas.
- Water the seedbed, if there's no rain, and try to keep the earth evenly moist so that there is no check to growth. Pull them when you like their size!

ALAN TITCHMARSH'S FAVOURITE CARROTS

- **Chantenay Red Cored** – A tried-and-tested old favourite
- **Early Nantes** – Reliable and good for successional sowing
- **Flyaway** – Sweet flavour and resistant to carrot fly
- **Sweet Candle** – Long, slender roots with good flavour
- **Parmex** – Short roots, good for containers and grow bags

'When my grown-up daughters were young they did not like carrots – until I acquired an allotment and grew my own. "But these aren't carrots!" they said, unable to believe that the sweet and tender roots they were eating were the same as those they had for lunch at school or bought from the supermarket. They proved the point that home-grown carrots have an infinitely better flavour than shop-bought.'
– **Alan Titchmarsh**

GOT A MORNING?

Plant up a summer salads container

Big harvests can come from small spaces, as lots of delicious vegetables grow well in pots. Many look very ornamental too, so if you plant them in a stylish container, they deserve pride of place. You'll have to water and feed them regularly, but this is easy to do if you keep them close to the house. And why have just one crop when you could have two?

You'll need

4 cordon cherry tomato plants: such as 'Sungold', 'Rosella', 'Gardener's Delight', 'Ildi'
12 lettuce plug plants such as 'Salad Bowl', 'Red Salad Bowl' and 'Green Oakleaf'
4 bamboo canes
Multipurpose peat-free compost
Large tub approx 80cm wide x 60cm deep x 30cm high, with drainage holes drilled in the base
Liquid tomato feed

Grow it

Cordon tomatoes are tall but take up little space at soil level, so why not plant lettuces at their feet? Both like a moist, rich compost, and the lettuces enjoy the light shade cast by the tomatoes. Stand the large tub in your sunniest, most sheltered spot. Plant the tomatoes 20cm apart, with a cane for support, then fill in around them with the lettuce plugs, 10cm apart. Water regularly. Once the flowers appear, start feeding fortnightly with a liquid tomato fertiliser.

Your harvest

Pick the tomatoes as they ripen in summer. With the lettuces, pick a few outer leaves from each plant every few weeks.

OTHER JOBS

Plant potatoes

Whether chitted or not, it's time to plant your potatoes. Dig out a trench that's deep enough to cover the top of your tubers with at least 5cm of soil. Sprinkle some general fertiliser along the base of the trench, then stand the tubers 30cm apart for earlies and 40cm apart for main crop varieties. Rake the soil back over and firm it down, leaving it slightly lower than ground level. This directs water to the newly growing potatoes until they emerge. Once the leaves start to appear, start pulling up more soil over them for protection against late frosts.

Look after chilli seedlings

Keep your chilli seedlings warm to encourage strong growth. When they are large enough to handle by the seed leaf, prick them out by lifting the seedling with a dibber and lowering the roots into pots of potting compost. Firm lightly and water them in but avoid drenching with cold water and put them back onto a warm windowsill or greenhouse bench. Give them plenty of space and add a cane to support the main stem as the branches develop. You can plant them outside in early June. Chillies produce the best crops in the sunniest spot of the veg patch and will keep cropping till autumn.

Sow summer cabbages

Cabbages have the advantage of being available fresh from the garden throughout the year. This month, spring cabbages should be ready for harvesting and it is time to start sowing summer and autumn varieties too. Sow into trays or modules of peat-free seed compost at a depth of 2cm and place on a sunny windowsill or in the greenhouse to germinate. Once seedlings are about five weeks old, harden off and plant outside into their final positions in the vegetable plot or allotment.

TRY SOMETHING NEW...

Achochas

This vine-grown vegetable, with a habit similar to courgettes, is a soft fruit that tastes a bit like cucumber when young and green pepper when more mature.

How to: Sow single seed 1cm deep in compost in small pots; water in and keep indoors. Once the plants have six leaves, remove carefully and pot up in bigger pots. Push a bamboo cane into each pot to provide support.

 # Rekha Mistry's Recipe of the Month

RHUBARB AND VANILLA CAKE

You'll need

23cm rectangular baking tin
175g butter, softened
250g golden caster sugar
3 eggs
1 tbsp vanilla extract
175g self-raising flour
300g rhubarb, cut into 2cm pieces and tossed with a little lemon juice
4 tbsp demerara sugar
Icing sugar to dust

Method

1 Grease and line a 23cm rectangular baking tin. You can also grease the paper and sprinkle it with demerara sugar if you like.
2 Beat butter and caster sugar, then beat in the eggs and vanilla. Fold in the self-raising flour, followed by the drained rhubarb.

3 Gently pour the batter into the tin, sprinkle generously with the demerara sugar and bake in a preheated oven at 180°C for 30–40 mins until golden brown and a skewer comes out clean.

4 Cool completely on a wire rack before dusting with icing sugar. The cake will keep for up to a week in an airtight container in a cool, dark cupboard.

 ## Wildlife Notes

Spring flowers such as crocuses, primroses and grape hyacinths are providing vital pollen and nectar for queen bumblebees, while the first of the nettles are sprouting in time for egg-laying peacock and small tortoiseshell butterflies. Keep an eye out, too, for early ladybirds – avoid the temptation to remove aphid clusters from plants, as they provide food for those further up the food chain. Blue tits and great tits are starting to nest around now, while house sparrows, robins and blackbirds may already be sitting on their eggs.

The warmer temperatures and rising of the sap are also encouraging moths to breed. They lay eggs on the newly burst leaves of native trees like hazel, hawthorn, birch and oak, plus long grasses and wildflowers that we gardeners typically think of as weeds. Those moth caterpillars are at the bottom of the food chain, and will go on to provide food for hundreds of hungry baby birds, plus hedgehogs, frogs, mice and voles. The more caterpillar food plants we grow in our gardens, the more wildlife there'll be.

The common toad is less common than the common frog, having declined by 68 per cent over the past 30 years.' – **Kate Bradbury**

LOOK OUT FOR...

- Toads making their way to their breeding ponds. Toads remain loyal to a breeding site for life, unlike frogs, which tend to be more opportunistic.
- Swifts and swallows. They don't typically appear until the beginning of May, but keep your eyes on the sky at the end of the month for the season's first individuals.
- Hedgehogs, which are now starting to emerge from hibernation. They will be feeling hungry, so leave a dish of chicken-flavour cat food and some water out for them.
- Red mason bees – the females lay eggs in hollow stems, then seal them with mud. If you have a bee hotel, you may spot them flying to and fro, carrying mud or pollen.
- Queen wasps – having emerged from hibernation, they visit flowers for nectar before scraping wood from trees, fences and benches to start a new 'paper' nest.
- Newts in your pond – they breed later than frogs and toads but will be busy feeding up on frog tadpoles to get into peak breeding condition.

WILDLIFE PROJECT: HOW TO HELP HEDGEHOGS

- **Create a highway** – Hedgehogs roam an average distance of 1–2km each night in their active season (March–November), so it's vital they can travel easily between gardens. Cutting 13 x 13cm holes in walls or fences will let hedgehogs through but be too small for most pets.
- **Make ponds safe** – Hedgehogs are adept swimmers, but if they can't climb out of steep-sided ponds or pools they will drown. Use a pile of stones, a piece of wood or some chicken wire to create a simple ramp.
- **Be untidy** – Leave plants unattended in a corner of your plot, and don't cut them back in winter – hedgehogs might nest here and will benefit from the abundant insects lured by the wildness. Use branches to add structure.

- **Clean up litter** – Hedgehogs can get tangled up and trapped in litter. Polystyrene cups, plastic and elastic bands are all common offenders. Also avoid using netting for crop protection. If you have to use netting, keep it taut and store inside when not in use.

- **Provide food and water** – Hedgehogs benefit from being given extra food to supplement their natural diets. They enjoy hedgehog food, meaty cat or dog food or chopped, unsalted peanuts. Water can also be scarce at certain times of the year and is the only thing you should give them to drink.

- **Avoid chemicals** – Lawn treatments reduce the populations of worms, while pesticides, insecticides and slug pellets kill off other invertebrates that hedgehogs feed on. A healthy, well-managed garden shouldn't need to rely on chemicals.

- **Check before strimming** – Hedgehogs will not run away from the sound of a mower or strimmer – check before you cut to avoid causing horrific injuries or death. Single hedgehogs are easily moved, but use gloves! Moving a hedgehog family is more complicated and ideally they should be left undisturbed – call the BHPS on 01584 890801 for advice.

- **Be careful with bonfires** – Piles of debris are irresistible to a hedgehog seeking shelter, so only build a bonfire on the day of burning or move the pile just before lighting it, to avoid a tragic end.

- **Build a log pile** – One of the best features for encouraging all kinds of wildlife – and so easy to make. It will encourage insects and provide somewhere for hedgehogs to nest.

Spotter's Guide to Bumblebees

It's no coincidence that bumblebees are furry. This helps them fly in the cold, so they can forage early in the year and first thing in the morning. Only mated queens (fertile females) survive the winter, hibernating in hedges or under logs. Each makes a nest from scratch, collecting pollen and nectar to feed the first grubs. The queen can't get help from these workers until they're reared to adulthood. Many bumblebee species are declining in the countryside because of meadow loss, but some thrive in gardens, where flower diversity is key to their success.

LOOK OUT FOR...

- **Buff-tailed** – Huge, black bee with brownish-yellow collar and waist bars, and dirty white, slightly orange or buff tail tip. Smaller, whiter-tailed workers appear at the end of the month. Nests in compost bins.
- **Red-tailed** – Another giant, but this time jet black and velvety, except for the stark, burning red tail. Buzzing feebly in the grass is not a sign of illness, just her way of shivering, using muscle heat to warm up.
- **Common carder** – A compact, medium-sized species that's bright brownish-orange all over, especially early in the year. Starts to appear in the middle of the month and often nests in large grass tussocks.
- **Tree** – Bright orange thorax, abdomen deep black, tail white. Recent colonist from Europe, most common in the south of Britain but spreading quickly, nests in tree holes and old bird boxes.
- **White-tailed** – Pure white tail and brighter yellow bars distinguish the queens from buff-tailed. Along with buff-tailed, it's one of the first to emerge in the year and is out from late March.

Troubleshooting Guide

DANDELIONS

Dandelions are weeds familiar to us all, but some of the fascinating facts about them may surprise you, not least their value to bees and role as a pollinator-supporting plant. Flowerheads appear from spring, each one consisting of hundreds of flowers, with every one of these maturing into a seed with a tiny parachute or 'pappus', which helps it stay airborne to travel far away from its parent. That way, it can thrive with less competition from its relatives. The leaf rosette can be large and upright (soaking up lots of sun for photosynthesis) or small and flat (so safe from being chopped off by mower blades). Down below, a taproot – thick, dominant and tough – is only destroyed if the whole of it is removed from the soil, as any bits left behind can regrow. Your best option is to dig far down into the soil, until you're sure you've removed every last bit of taproot. Try to avoid the use of chemicals, but if using them as a last resort make sure to never spray when the flower is open, as this is when they are most attractive to bees and other pollinators.

TOP TIP

If you have no time to dig out whole roots, at least cut off the flower-heads before they produce seeds.

CONTROLLING GROWTH

All plants want to grow, usually upwards to compete for light. Young plants put their energy into one or a few upright shoots, rather than growing sideways. These shoots exhibit apical dominance – producing a chemical in their tips called auxin that suppresses sideshoots. Watch this happening in saplings, which have tall main stems pushing towards the sky and very little lateral (sideways) growth.

If you prune off the tip of a dominant shoot, you will notice that the sideshoots start growing – you have stopped the auxin production that inhibited sideshoot development. So, to create a nicely shaped lavender, trim off the shoot tips now, encouraging the plant to make sideshoots over the next few months and preventing it from becoming leggy. With some plants, particularly trees, you want a single shoot to lead the way upwards, producing a tall, elegant shape with a trunk. If you have a tree or shrub you're happy to experiment with, have a go at creating different growth effects.

PRICKING OUT

Pricking out is a satisfying job that will give you thriving seedlings, more robust than any you have left behind in the seed tray. When you scatter seeds, some germinate too close together, others not at all. To make sure you get robust seedlings that will develop into healthy plants, when pricking out, choose the strongest seedlings and give them more space in a deeper pot of compost. Seeds don't need rich compost to germinate, but your seedlings will do better in a compost with added feed. This way you save on compost and plant food, plus space, too, by reserving larger pots for the favoured few.

With fast-growing seeds such as squashes, it's worth sowing straight into a module or pot, which they will soon fill. The whole thing can simply be repotted or planted out once the roots have reached the edges.

ROSEMARY BEETLES

These beetles can be a menace. It is worth remembering that as well as rosemary and lavender, these beetles and their larvae will also damage sage and other closely related plants, so check those, too. On the whole, the beetle rarely does enough damage to have a significant impact on the plants' health, though. Try to collect as many beetles as possible by hand – you can speed things up by spreading an old sheet beneath the infested plants and then shaking them. This way, you can collect signif-icant quantities of these pests. There are a number of natural predators of the rosemary beetle including ground beetles (which eat the larvae), frogs and toads, so ensuring that your garden is a good place for these creatures to live will also mean that you have plenty of help.

POTS FOR SHADE

There are lots of gorgeous plants to choose from for containers in a shady spot: Azaleas and other rhododendrons and the shrubs fothergilla and mahonia would all work, or some herbaceous perennials such as pulmonarias, primulas, polyanthus, tiarella, heucheras, epimediums and hellebores. Between them, they will not only provide pretty flowers, but in many cases will also have gorgeous foliage, so provide a pretty display over a much longer period than flowers alone. Used to increase the length of the main display and add welcome pops of colour.

MAY

May is to be relished for all its outrageous glory at every moment of every day. Perhaps the best way to enjoy the garden this month is to simply get outside. Give your garden your time, even for the most humble of jobs, and in May it will give you heaven in return. The weather can of course be fickle, and it is a rare May that does not have a week or more of grey, wet weather. Frost is still a real risk, too, but this never spoils the Maytime party.

The soil has warmed up and the days are long – it's light by five in the morning and, by the end of the month, it will still be light enough to garden outside at 10pm, so plants are growing strongly. The vegetable garden is still a bit limited in May – winter's harvest is finished, and summer has barely begun – but fast-growing salad leaves, like rocket and lettuces of all kinds, are at their very best. And there's plenty of sowing and planting to get on with now the soil has warmed up.

There is one major problem with the month of May that no gardener can fix: it has only 31 days. Other than that glaring shortcoming, it is almost perfect. Everything that makes life worth the struggle comes gloriously into being in May. These are the days of blossom and a growing, swelling blowsiness of hedge and border. The grass is emerald green as spring shifts easily into summer.

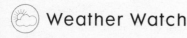 Weather Watch

May is when the growing season gets into its stride. Frosts this month are relatively rare, but the risk is greater the further north and the higher up your garden is. It takes only one frosty night to cause havoc among tender plants, so pay close attention to the forecast. Easterly winds become more common in May as the prevailing westerlies from the Atlantic tend to weaken.

This means spring can often be put on hold in eastern areas as a cold, grey blanket spreads from the North Sea, lasting several days and slowing growth. But the opposite is true for sheltered western areas, especially to the west of high ground, which acts as a barrier. May is on average sunnier than most of the summer months in places like north-west Scotland.

WEATHER FACTS

- Average temperature in the UK: 14.7°C
- Highest number of days of air frost: 1.9 in northern Scotland
- Highest average number of hours of sun: 216.3 in the south east of England
- The sun is as strong in May as it is in July.
- Average days of rainfall this month: 8.8 in London, 11.3 in the south west of England, 14.3 in northern Scotland and 12.9 in Northern Ireland
- Plant growth begins at 7°C. The average minimum temperature this month is 6°C, so this is a great month to be planting.
- May is the peak for sunshine hours. The average is 186 hours of sunshine. This decreases throughout the year due to cloud cover.
- This is the month of the year with the least rainfall across the UK.

WEATHER ALERT

Strong sunshine in May can lead to big temperature fluctuations between day and night, which is magnified by a greenhouse or cold frame. Be sure to open vents on sunny days, but don't forget to close them in the evening, especially if late frosts or chilly east winds are forecast.

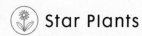

Star Plants

10 OF THE BEST

1. Aquilegias

Once you have one aquilegia, they will always be with you – they are big self-seeders. The name comes from the Latin for eagle – the long spurs at the back resemble the talons of an eagle. Try the popular 'Nora Barlow', which has been wooing admirers for more than 100 years. Its distinctive double flowers are useful for early summer colour. Flowers May to June.

2. Alliums

The allium family is extraordinarily varied. It encompasses not only the humble onion and unpretentious leek, but also all manner of ornamental bulbs in an enormous range of shapes (from dangling bells to sputniks the size of melons) and colours (from deepest purple to cleanest white). Try *Allium* 'Purple Sensation', which has huge purple flowers the size of tennis balls. Flowers May to June.

3. Geums

The humble geum is tough, long-flowering, fuss free and comes in a range of beautiful shades. One of the most well known is the red 'Mrs J. Bradshaw'. Depending on variety and area, geums can flower from May to September.

4. *Magnolia wilsonii*

Discovered in China by the great plant hunter E.H. Wilson, this is a shrub of great beauty. Its large pendulous flowers have a mulberry-coloured centre as grand as a papal tiara and petals the colour and texture of freshly laundered fine cotton sheets. Best of all they're wonderfully scented. Flowers May to June.

5. Lilac

These classic garden plants can be shrubs or small trees and have fragrant tubular flowers from late spring to early summer which are attractive to insects. Try *Syringa vulgaris* 'Primrose', which has white flowers with a strong scent. Flowers May to June.

6. Spring-flowering clematis

These climbers are ideal for covering a fence, pergola or bare wall and are at their best this month. Try the 'Frances Rivis', which has a prolific display of large, blue to purple flowers. Flowers April to June, depending on variety.

7. Solomon's seal

This graceful perennial is an old favourite of cottage gardens and a lovely choice for a shady spot with moist humus-rich soil. The bell-shaped, white flowers dangle from arching stems. Flowers May to June.

8. Choisya

A popular evergreen shrub that will add easy structure to a border. Try the compact 'Aztec Pearl', which has clusters of scented white flowers. Flowers April to May.

9. Hardy geraniums

Commonly known as cranesbill geraniums, these tough plants are reliable, long flowering and easy to grow. They're attractive to pollinators and come in a huge range of colours, with many thriving in partial shade. Try the violet blue 'Rozanne', which flowers for months. Flowers May to October depending on variety.

10. Bearded iris

Tall bearded irises are not only the most flamboyant perennials we can grow, but they're also rare among plants in being available in every single rainbow colour, not to mention myriad colour combinations, including black. The display is fleeting but worth it. Try 'Black Swan', a sultry rich-purple and black variety. Flowers May to June.

'If we think of our gardens as natural habitats, we can gather information about which plants are likely to thrive and how best to grow them. The lesson here is to garden with nature rather than against it. Once we realise this, then we can embrace the conditions we're presented with, instead of seeing them as problems.' – **Carol Klein**

WHY NOT TRY...

Filling your garden with scent

Scented superstars fill the month of May with a cocktail of swooning fragrances and sumptuous blooms. Lilacs and daphnes compete for perfumed prowess, flooding the garden with their floral tones. Others, such as lily of the valley and *Akebia quinata*, are more elusive and only reveal their aroma up close and personal. The best plants for scent are:

- Lilac 'Katherine Havemeyer'
- *Osmanthus delavayi*
- Lily of the valley
- Chocolate vine (*Akebia quinata*)
- Garland flower (*Daphne cneorum*)

WHY NOT TRY...

Best blossoms

All the exuberance of the month is captured by ornamental cherries as their bare branches are transformed by clouds of confetti-like blossom. Crab apples and May hawthorns also take part in this floral overdrive, as does the Judas tree, with its profusion of tiny sweet pea-like flowers wreathing its stems. To sit beneath any of these trees is a magical Maytime treat. The best plants for blossom are:

- Japanese crab apple
- Judas tree
- Ornamental cherry 'Pink Shell'

- Double pink hawthorn
- Apple 'Arthur Turner'

NOW'S THE TIME TO...

Plant out gladioli

Now's the time to plant out gladioli, both the big showy ones and the lovely delicate-looking and fragrant acidanthera (*Gladiolus callianthus*). These bulbs – or strictly speaking corms – are great cut-and-come-again plants.

If you have heavy soil, the corms may rot if planted directly outside. Instead, plant in pots (six corms to a 2-litre pot) from late March, and keep them well watered in a bright, frost-free place. For a succession of flowers, plant regularly until late June. Then plant them outside in clumps when about 30cm tall, after the last frost. Add grit to the planting hole to improve drainage.

GLADIOLI TO TRY

- **'Green Star'** – Most green gladioli seem to fade to a hideous urine yellow, but not 'Green Star', which stays miraculously bright acid right through its flowering cycle, even when the blooms start to go over. It reaches 1.2m tall.
- **'Espresso'** – With its rich, crimson-black flowers, 'Espresso' is a striking variety, but it can be hard to get hold of. A good substitute would be 'Sweet Shadow', which has large maroon-black flowers. It's quite tall, reaching 1.3m.
- **'Plum Tart'** – A great choice for the cutting patch, 'Plum Tart' produces flower spikes packed with velvety, purple blooms. Plant it in succession and you'll get flowers from midsummer through to autumn on spires up to 1.2m tall.

🧭 Jobs to Do this Month

The gardening to-do list in May is a long one, but it's exciting to be out in warm weather, getting things under control and transforming the garden for the seasons ahead. As well as tying in climbers, cutting back shrubs, sorting out the lawn – all those maintenance tasks, there's still time to get new plants into the ground, sow annuals and plant up plugs for pots and hanging baskets. You could plant up a summer pot or choose some new clematis to cover boundaries. This is a month full of promise.

KEY TASKS

- Mow the lawn once a week and feed.
- Harden off tender plants.
- Thin out seedlings.
- Prune spring-flowering shrubs.
- Sow annuals.
- Sow hardy veg direct.
- Trim hedges.
- Look out for pests such as greenfly on broad beans.
- Transplant greenhouse tomatoes.
- Remove weeds from ponds.
- Keep on top of weeds before they get too big.
- Water blueberries with rain, not tap water.
- Go on a slug hunt before they have time to multiply.
- Give fruit trees in pots a liquid feed for a boost.
- Sow catch crops in any gaps in your plot.
- Pull out new raspberry canes that are sprouting outside rows.

GOT TEN MINUTES?

Grow more herbs for free

Pots of herbs sold in shops are usually grown from seed. Each pot contains a mass of seedlings. Save money on buying herbs by splitting them up and growing them on to get a supply for the rest of the summer. These can be placed on a sunny, warm windowsill or planted outdoors in pots or in the herb garden after the risk of frost has passed. Basil, coriander and dill will be great all summer, but chives, mint and sage can take a permanent spot in the garden.

1 Support the plants with fingers widespread before turning the pot upside down and gently knocking out the rootball.
2 Pull the roots apart to give several small clumps of seedlings. Trim the base of the roots by a third if long.
3 Plant each clump into a new pot and firm in, to the same level that they were in the previous pot.
4 Trim the upper foliage to encourage new growth. Soak the pots and space them out well for maximum growth.

GOT AN HOUR?

Take bedding plant cuttings

Bedding plants like verbenas root easily from cuttings, especially now. They can be placed in a tray on the windowsill with a clear polythene bag as a cover or kept in a propagator in the greenhouse. These cuttings are known as softwood as they are taken from the soft new growth at the tips of shoots. This kind of growth forms roots easily so it's worth having a go as this is a cheap way to make more plants. Once they've formed roots at the base, they can be potted into individual pots to grow on for free bedding plants next year.

1 Choose strong non-flowering shoots from a healthily growing plant. Reduce them to 8–10cm long.
2 Remove the lower leaves, leaving the top ones. Pinch out the soft tip and trim the base just below a node.
3 Make a hole in the compost with a dibber and insert the cutting. Firm it in, then soak thoroughly with tepid water.

GOT A MORNING?

Growing from plug plants

There is something comforting about growing plants from tiny plugs. The tricky part of germination is taken care of and you have a collection of mini plants that stand every chance of succeeding if you treat them well. Plug plants are seedlings that are a few weeks old, with well-established roots, growing in small cubes of compost, meaning they can be transplanted to a wider spacing in a more compost with little shock to their system. They're available online and from seed merchants, so there's plenty of choice and they vary in size from those the size of a small flowerpot, to those as small as the end of your little finger. They are useful at this time of year when they can be planted out in just a few weeks' time. Just make sure the plugs never dry out completely and you are assured of success.

1 Unpack and inspect the plants as soon as they arrive – they should look strong and healthy. Place them in good light and keep moist in a tray of water.
2 Transplant the plugs as soon as you can into small pots of peat-free multipurpose compost or trays of the same. The plants should be given room to expand.
3 Fill in with compost around the roots and lightly firm into place. Alternatively, tiny plugs can be spaced out and planted in trays using a dibber of an appropriate size.
4 Stand the newly transplanted plugs in a bright, frost-free place and keep moist. Tiny plugs will need to be potted on. After four to five weeks they can be hardened off.

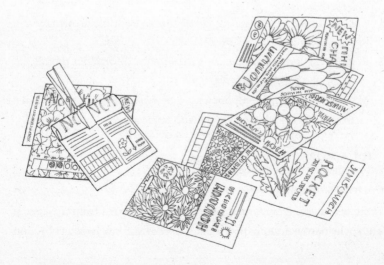

STEPS TO SUCCESS

- **Watering is key** – as they grow the plants' moisture requirements will increase. Feel the compost with your finger on a regular basis and give it a good soak whenever it feels slightly dry. There's no need to keep it sopping wet, but an even supply of moisture will make for even growth.

- **Watch out** for greenfly, and wash any off before they have a chance to multiply.

- **Keep the plants** well ventilated, especially if you're out all day, so they don't cook in scorching sunshine. On really bright days, a sheet of newspaper or old net curtain draped over them will help to prevent them from crisping up. Horticultural fleece may exacerbate the problem, as it doesn't allow much in the way of air circulation.

- **Pinch out** any shoots that grow tall and leggy, to encourage bushiness, although many bedding plants branch naturally and won't need such attention.

- **Harden off** your plants before moving them outside permanently. This simply means standing them outdoors during the day and bringing them back under cover at night. Ideally, do this for a week or more to accustom them to the lower temperatures they'll encounter outdoors. At the very least, make sure they've had plenty of fresh air. Water them just before planting them in the ground or into a container.

- **Plant them** in their final location at a spacing that will allow them to grow to their full size. Also take steps to protect them from slugs and snails, which will appreciate their tender young foliage, even when they have grown apace.

AND A FEW OTHER JOBS...

Extend flowering with the Chelsea chop

Cut back herbaceous perennials such as echinacea, helenium, aster and anthemis in late May (around the time the Chelsea Flower Show is usually held) to get more blooms later in the flowering season. You can prune all the stems in a clump to delay flowering for four to six weeks. Alternatively, cut back around half of the clump now, so flowering takes place over a longer period. Plants for the Chelsea chop include:

- Phlox
- Achillea
- Campanulas
- Asters
- Echinacea
- Rudbeckias
- Sedums
- Penstemons

Take stem cuttings

Exploit the new flush of growth on penstemons, achilleas and phlox to make new plants. Sever shoots about 8cm long, nip off bottom leaves and growing tips and dibble in around the side of a pot, in well-drained compost.

Prick out seedlings

Seedlings grow fast at this time of year. Prick them out now into pots or modular trays to give them a boost of plant food in the fresh compost and more room for the roots to grow. Water them in as soon as they are pricked out and keep spacing them out as they grow into strong, sturdy plants.

- Knock the pot on the bench to loosen the compost then gently tip the seedlings out of the pot, avoiding breaking them.
- Pick up the seedlings by the leaf between thumb and index finger to avoid damaging the fragile stems.
- Fill pots with multipurpose compost, firm the surface, then make a hole in the centre and plant the seedlings. Firm again.

Plant out sweet peas

If it wasn't possible in April, plant sweet peas now and they'll be flowering in just a few weeks. These are hungry feeders, so plant in a richly prepared piece of ground or pot and have the climbing supports – such as a tepee or pea netting on a frame – built ready for them. Plant a small clump for a mass of flowers or put single plants at the base of each support. Nip the top out of individual plants to encourage sideshoots, unless you plan on training single stems to get just a few quality blooms ready for a special indoor display. For the best results, keep sweet peas well watered and add liquid feed to the water once a week. Pick as soon as the blooms open to keep the plants flowering.

Plant up a summer pot

Large plant-filled containers can make a big impact in the garden. Larger plant pots retain moisture longer, remain upright in strong winds, thanks to their weight, and provide more sustainable nutrition to their inmates, giving far greater impact than a few smaller containers. Plant up your pots for a dramatic show – look for thrillers, fillers and spillers – the latter to cascade over the edge. Drainage is key, so don't block the hole in the base of the pot. I use an equal-parts mix of multipurpose compost and John Innes No.2, the former to prevent the compost becoming claggy, and the latter for weight and its ability to retain nutrients. Apply dilute liquid tomato feed every couple of weeks after the first month. Deadhead plants regularly and rotate the pot occasionally for even growth.

How to best create your seasonal display:

1 Place the pot in its final location before filling it with compost. Lightly firm the compost, and position any supports before adding the plants.

2 Thoroughly water all the
plants before taking them
out of their pots – dry
rootballs will remain dry
and the plants will then
shrivel if not properly
soaked at the outset.

3 Plant any climbers at the
foot of the support system
and carefully tie in the stems
with soft twine so that they
are encouraged to climb.

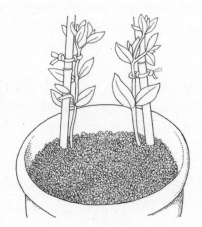

4 Add bushy plants, with
trailers cascading over the
edge. Leave no more than
10cm between plants so
that they'll fill the whole
pot. Water in well.

Cover boundaries with clematis

Clematis are among the most spectacular climbers in the garden – up walls, clambering through large shrubs or spiralling up elegant obelisks – where would we be without clematis? And there are so many kinds – from the rampant *Clematis montana* to the herbaceous types that sit happily in beds and borders, and the large-flowered hybrids that cause onlookers to gasp at their blooms. And, of course, the hybrids of *C. texensis* and *C. viticella* – the first with slashed goblet-shaped flowers, and the second with elegant pixie hats that have grace, as well as a resistance to dreaded clematis wilt. Late summer and autumn are decorated by the yellow bells, followed by the seed heads of *C. orientalis* and *C. tangutica*. Planted now, they'll romp away to flower in summer, and some varieties will bloom from spring until the start of autumn.

How to plant clematis:

1 Plant clematis a little deeper than it was in its container. All types benefit from this – it helps to keep their roots cool and can also mitigate against clematis wilt.
2 Sprinkle a general fertiliser, such as blood, fish and bonemeal, in the hole to offer nourishment. Firm the soil, taking care not to fracture the stems.
3 Place a couple of pieces of broken paving over the soil to prevent the earth drying too rapidly. Most clematis like their heads in the sun and their roots in cool earth.
4 Give the plant a good soak after planting and ensure it has a support system around which its tendrils can curl. Attach stems to the support with soft twine.

TOP CONTAINER PLANTS

- *Begonia* 'Apricot Shades' — Wonderful lasting colour
- **Bidens** — A yellow-flowered trailer to tumble over the rim
- **Clematis** — Great for climbing up a tripod and there are plenty of varieties to choose from
- **Scented-leaved pelargoniums** — Aromatic foliage
- *Salvia* 'Purple Majesty' — Richly coloured flower spikes

'We gardeners often find ourselves wishing for just a little more growing space to fit in a few more plants. But making the most of what you've already got is key. By re-evaluating your garden or outside space, you will soon see that there are many places where you can squeeze in extra colour, scent and edible plants. Walls, fences, window ledges and the side of a shed are all areas where you can add plants.' – **Arit Anderson**

ALAN TITCHMARSH'S FAVOURITE CLEMATISES

- *C. viticella* **'Purpurea Plena Elegans'** – Purple double blooms
- *C.* **'Étoile Rose'** – Slashed bells of deep and pale pink
- *C. tangutica* – Bright-yellow waxy bells, then fluffy seed heads
- *C.* **'Perle d'Azur'** – Spectacular, wide-faced, pale-blue flowers
- *C. montana rubens* – A profusion of pink blooms in spring

DID YOU KNOW?

You can eat young, fresh nettle tips (yes, really), they're a magnet to all sorts of beneficial wildlife, they make great plant food, and they're just the thing for kicking your compost heap into action. And they're free. What more could you ask for in a plant? The nettle we're used to in this country, *Urtica dioica*, is a perennial plant crammed full of iron, calcium and magnesium – handy if you're looking for a superfood. It's also full of nitrogen, which is why it is so good for plant food and compost heaps. OK, nettles can give a bit of a sting, but the benefits certainly outweigh this if you handle with care.

5 THINGS TO DO WITH NETTLES

1. Make a soup
2. Turn them into liquid plant food
3. Eat them as spring greens
4. Add them to the compost to speed up decomposition
5. Leave a patch growing in your garden for butterflies to lay eggs on

Time to Prune

May is the month to prune the less hardy shrubs and plants in your garden. If they've survived the worst of the cold, their buds should be breaking by now, and cutting the plant back will tidy the shape and encourage vigorous growth. While the hardiest plants are tough enough to be pruned in winter and early spring, half-hardy and tender plants are best left until temperatures reliably stay above freezing overnight. This will vary depending on where in the UK you garden: the milder south and west are usually frost-free by early May; the east and north by the end of the month. And your location will also dictate which plants are reliably hardy: in the north east, pittosporum and escallonia, for example, are best pruned now – a month or so later than in the south west where the weather is much milder.

PLANTS TO PRUNE

- Silver wattle
- Callistemon
- Cistus
- Hibiscus
- Lonicera (evergreen)
- Olearia
- Penstemon
- Phlomis
- Sage
- Santolina

In the Greenhouse

May in the greenhouse is a busy time. Seedlings are growing fast, tender plants are waiting to be hardened off and moved outside and the temperature is increasing, which means shading might need to be put up on sunny days to protect plants. Light conditions are high now and plants are growing fast, so be efficient with the watering. Only water directly onto the compost surface. Splashes on the leaves make them vulnerable to scorch on hot days so avoid this unless they are well shaded. There can be problems with fungal rot if you water in the morning, leaving a lower humidity in the greenhouse at night. Always fill watering cans ready for the next watering as this allows the water temperature to rise to the ambient temperature in the greenhouse, meaning the plants suffer less stress.

JOBS TO DO THIS MONTH

- Cordon (vine) tomatoes are unruly sprawlers that must be trained into single-stemmed fruiting plants. Sideshoots appear in the axils between the leaves and main stem. Pinch them out between your thumb and forefinger when they are tiny – do this weekly. Avoid the growing tip at the top of the plant and leave it to climb up the support.
- Damp down by spraying water on the floor to increase humidity, which discourages red spider mite infestations.
- Finish planting up hanging baskets.
- Put up shading to avoid plants becoming scorched and to lower the temperatures during the day.
- Keep pricking out and potting on seedlings and young plants to give their roots room to grow.

DON'T FORGET TO...

- Cut out green shoots from variegated shrubs.
- Add new ponds plants like water lilies.
- Water newly planted trees and shrubs.

- Harden off tender plants for a summer outside.
- Plant up hanging baskets.
- Repair damaged lawns.
- Plant out dahlia tubers once frosts have passed.
- Water early and late in the day to make the most of the water.

LAST CHANCE TO...

- Plant summer bulbs.
- Sow tender veg such as melon, sweetcorn and cucumbers.

 # Fresh from the Garden

May marks the start of the big push, when fruit and veg planting really begins and seed sowing accelerates. It's time to plant out young tomato plants sown in February – with a bit of caution, as even in the south there can still be late frosts. Although daytime temperatures are rising, the nights can still be quite chilly, so newly planted outdoor vegetables need a little protection. Cover them all with horticultural fleece every evening, for the first few weeks after planting out.

Runner beans can be planted out, and grown again using supports such as beanpoles. And it's also time to get the main potato varieties into the ground – good reliable varieties to try are 'King Edward' and 'Désirée'.

Once the veg plants are in, a great job to do is make organic plant feed – comfrey tea is an easy one. Simply fill a bucket full of comfrey leaves and top up with rainwater, then secure with a lid. Set this aside to brew for the next four weeks when it will be ready to dilute and use as tomato feed.

WHAT TO SOW

- Beetroot
- Cauliflower
- Peas
- Spinach

- Chard
- Carrots
- Chard
- Chicory
- French and runner beans
- Climbing beans

WHAT TO PLANT

- Hardened off seedling of cabbages and leeks
- Broad beans
- Potatoes
- Peppers and chillies
- Salad leaves
- Tomatoes
- Runner beans
- Strawberries

WHAT TO HARVEST

- Lettuce
- Asparagus
- Chard
- Salad leaves
- Turnips
- Broad beans
- Radish
- Rhubarb

 On the Veg Patch

KEY TASKS

- Pinch out the growing tips on broad beans.
- Finish earthing up potatoes to protect from frost and encourage a bigger crop.

- Harden off and transplant well developed brassicas, celery and leeks.
- Sow hardy annual herbs outdoors.
- Sow cauliflowers, carrots, French and runner beans and parsnips outdoors.
- Weed between outdoor seedlings.
- Prick out seedlings sown into seed trays and pot on into 9cm pots. Grow on before hardening off and planting outdoors.
- Sow batches of peas in small pots to provide shoots for salads.
- Support larger pea plants with pea sticks.
- Sow a second batch of fast-growing veg to provide a steady supply.
- Pot on tomatoes, chillies and peppers into large containers. Insert canes and tie in so that they have some support as they grow.
- Protect kale and carrots from pests by covering pots with fine mesh or fleece.

CROP OF THE MONTH: ASPARAGUS

Did you know? – It used to be believed that if a pregnant woman ate asparagus she was more likely to give birth to a boy. While this isn't true, asparagus is beneficial to eat during pregnancy due to its high folic acid content.

Nutrients – It is rich in folic acid (vitamin B9), which helps the body produce healthy cells, and vitamins A, C, K and B6. A powerful diuretic.

Storing – Will keep for a few days in the fridge if placed upright in a jar with a little water. The spears can be blanched for 5 minutes and kept in the freezer for longer-term storage.

Good cultivars – 'Backlim' is an all-male plant producing good yields of fat spears, while 'Pacific 2000' produces tender, stringless, green spears.

HERB OF THE MONTH: DILL

An attractive annual herb with delicate, feathery foliage that is easy to grow from seed. Choose a compact variety and you can grow it on your windowsill. Dill can be added to salads, and is ideal with fish and eggs. Try the acid-green flowers, too, which are popular in Scandinavian

cooking. Grow dill in a pot at least 15cm wide. Recommended varieties: 'Bouquet' and 'Diana'.

GOT FIVE MINUTES?

Thin out carrot seedlings

Thin carrots to avoid overcrowding, leaving 10cm between plants. To reduce the risk of attracting carrot fly, water the drill before and after thinning, and pinch them out rather than pulling, which releases more scent from leaves. Thinning in the evening also helps prevent carrot fly problems. Don't waste the thinnings – they can be used raw in salads.

GOT 15 MINUTES?

Tend to strawberries

Prepare your strawberry plants now, ready for a great harvest next month. Start by weeding and removing all the oldest leaves at the base. This disturbance will attract birds to come and clear grubs and slug eggs, so leave them to it for a day or two. Next, spread a general fertiliser on the soil surface and put in slug traps or organic pellets for extra protection. Finish by tucking fresh straw under the plants to keep the fruit clean, then protect them with netting to keep the birds from eating the fruit.

GOT AN HOUR?

Pot on tomatoes

It's time to get your tomatoes into their final pots, ready for cropping. They can go outside this month once the risks of frost are over in your area. You can use containers inserted into growing bags so the roots can spread deeply and the plants get lots of nutrients. Put in strong stakes and tie the plants onto them to support the weight of the crop.

'My heart is racing as the outdoor air warms up and the seedlings start shooting up day after day. It seems like there is too much to do, but the reward will be a plot filled with new growth.' – **Rekha Mistry**

OTHER JOBS

Harden off veg plants

Young plants grown indoors or bought from the garden centre need to get used to being outside before they are planted in the ground. Tender annual veg such as chillies, courgettes, salads and tomatoes, herbs like basil and annual cut flowers need gradual acclimatisation. Over two weeks, introduce them to the conditions of their final planting place. In the first week, place them outside in a cold frame or sheltered spot during the day, then bring them in at night. The next week, leave them out and cover them at night, then leave them uncovered.

Control blackfly

Blackfly infestation on broad beans is inevitable at this time of the year as they tend to colonise the new soft, sweet tissue on the shoots at the top of the plants. The best way to stop these pests from getting a hold is to pinch out the tips of every plant as soon as you spot them. All the pods full of beans are set lower down on the plants, so you'll be able to look forward to a clean harvest later in the year.

Sow sweetcorn outside

Choose a warm, bright site to grow sun-loving sweetcorn, forking plenty of compost deeply into the soil to lock in moisture. Once the soil has warmed up to 10°C, sow seeds 5cm deep in a grid or block and space the plants about 45cm apart in each direction to assist pollination. Seeds can also be sown in small pots under cover now, and planted out once they are growing strongly.

HARVESTING TIPS

- **Leafy veg** – Salads, kale, herbs and the leaves of beetroot should be picked by hand or cut with scissors. Harvest a little from each plant, leaving plenty of leaves behind so that the plant can continue to grow. You can employ the same technique with lettuce – rather than harvesting a whole head, just pick the outer leaves and leave the rest to keep growing.
- **Broad beans** – Harvest broad bean pods starting at the bottom of the plant – another layer of beans will mature above them, so cut them off carefully to avoid snapping the stems.
- **Asparagus** – Asparagus takes a couple of years of nurture and patience before the plants are strong enough for you to begin harvesting the spears. Gradually increase the harvest period until the plants are well established. Cut when they are 25cm high, using a sharp knife plunged just below soil level. You can continue harvesting on established beds until the end of June, after which the plants need a good top dressing of fertiliser and a long period of recovery if they're to keep up the supply of spears for next year's harvest.

How to Grow Veg in Containers

If you have limited space or just want to grow more veg, planting crops in a container is the ideal solution. At this time of year, veg plants will grow rapidly so they'll mature to harvest stage before the end of summer. This is also a great way to have a go at veg growing if you've never tried it before.

TOP TIPS FOR GROWING IN POTS

- **Compost** – With fast-growing veg, such as salad leaves, there is no need to empty all of the compost out of the pot before re-sowing – simply replace the top third with some new compost.

- **Pests** – Check plants regularly to stop pests becoming a problem. Look on the undersides of leaves and under rims of pots where they could be hiding. Most pests can be removed by hand or blasted with a jet of water.

- **Watering** – Give pots a thorough soaking, rather than just a sprinkle, and direct the water to the base of the plants. Remember that it's not just warm weather that dries out containers – windy weather will wick away moisture, too.

- **Feeding** – Multipurpose compost will provide enough nutrients for about six weeks. After that, container-grown veg will be reliant on you for their food. Crops that bear fruit need potassium – try comfrey or tomato feeds. Leafy crops require nitrogen – make a homemade nettle fertiliser or use pelleted chicken manure. Seaweed feed is an excellent all-round fertiliser. Generally, feed every week or so, but follow the instructions on the pack. Don't be tempted to overfeed as this can lead to problems.

- **Potting** – Use the biggest container you can find as it will make watering and feeding the plants much easier. Ideally place it where it'll get sunshine for part of the day. Keep the compost just damp, feed with vegetable liquid fertiliser every fortnight, protect from slugs and keep birds off. Finally, with many vegetables, the more you pick, the more you'll get, so don't forget to keep harvesting.

VEG TO GROW IN POTS

- Salad leaves
- Beetroot
- Bush tomatoes
- French beans
- Round carrots
- Herbs
- Chillies
- Compact courgettes
- Cucumbers

FRUIT TO GROW IN POTS

- **Apple** – Grow on an M26 rootstock in a large tub
- **Fig** – The restricted root run will encourage fruiting
- **Lemon** – Give them frost-free shelter in winter
- **Blueberry** – Plant in ericaceous compost
- **Strawberry** – Undemanding, but remember to water them!

TRY SOMETHING NEW...

Japanese squash 'Uchiki Kuri'

A creeping vine producing orange, globe-shaped fruit.

How to: Sow individual seeds lengthways into 7cm pots filled with multipurpose compost. Plant out in summer. Once in flower the plants will respond well to a weekly organic feed. Harvest in late autumn. Fruits will store well for up to three months.

TRY SOMETHING NEW...

Welsh spring onions (*Allium fistulosum*)

These are a perennial variety, with strong, upright, green leaves. I also grow a variety with red bulbs. Perfect for winter salad lunch bowls.

Because they love cool conditions, they are ideal for growing over winter, as well as starting from new seed in late winter for fresh early-summer growth.

 ## Wildlife Notes

May is a noisy, busy and exciting time for wildlife. Many species are emerging from hibernation, while some, such as slow worms, have started breeding. As well as the dawn chorus reaching its peak in the first week of May, you may also be woken by snuffling and grunting hedgehogs at night. Mating is high on hedgehogs' agenda this month, and competing males may fight in a stag-like rut, which involves head-butting and occasionally pushing the rolled-up rival male away. Elsewhere in the garden, bees are buzzing and leaves are dotted with the eggs of butterflies, moths, shield bugs, leaf miners and other insects. Be careful not to remove aphids, caterpillars and other garden 'pests' from leaves, as these are all food for hungry mouths and beaks further up the food chain.

WILDLIFE WATCH

Check hedges and shrubs for birds' nests before pruning.

LOOK OUT FOR...

- Queen wasps – these have emerged from their winter torpor and are starting to nest, with the first batch of workers rasping wood from fence panels and garden furniture.
- The grass snake, which can grow to over a metre, and is olive-green in colour with a yellow and black collar, paler belly and dark spots down its sides. Look out for them in long grass and compost heaps.
- Hedgehogs, which are typically solitary, courting and mating.
- Broad-bodied chasers – these dragonflies start emerging from ponds this month. The females are brown, while the males are a distinctive powdery blue.

- Bats – now's a good time to spot them, as they're currently very active. Watch for them against a backdrop such as a row of trees, large hedgerow or open water.
- Flower beetles – unmistakeable and widespread, males have a vivid green body and fat hind legs.
- Leafcutter bees making semi-circular incisions in the leaves of roses and other plants. They chew out the perfectly formed discs to make their cigar-shaped nests in the soil.
- Damselflies, the first of which are flapping around ponds. Blue species are more common, but look out for the red ones, which are often earliest on the wing.
- Male adela moths, with their huge, eye-catching antennae and dark, metallic-green wings spread wide as they flutter together in the sunshine.
- Grounded bees on chilly mornings, buzzing to warm up for flight. They're merely shivering to generate sufficient body heat to get their muscles working fast enough for full flight.
- Comma butterflies – the new generation is starting to emerge, wings fresh and brightly coloured despite their ragged edges.
- Female orange-tip butterflies, which lack the orange tips, flying around, laying eggs on garden honesty. The males are less in evidence at this time of year.
- Swallows and swifts returning from southern Africa, filling the skies with their twittering song. These familiar summer birds are commonest in rural areas.
- Sparrowhawks – these classic garden predators are spectacular to watch, with their agile flight and powerful talons, as they chase and catch small birds.
- Blue tits. Parents will be frantically feeding their young, now. Let weeds flourish at the back of borders so moths can lay eggs on them and the tits can feed the caterpillars to their young.
- Kestrels. Familiar to motorway users, kestrels can also be seen over gardens, hovering characteristically as they hunt.
- House martins building their nests out of tiny balls of mud. Now, sadly, in decline, these beautiful blue-and-white birds nest in noisy colonies.

DID YOU KNOW?

- International Dawn Chorus Day is held on the first Sunday of May, because birdsong is at its loudest and most impressive at this time. All over the world, people get up early to listen to birdsong.
- Most ladybirds are carnivorous, and an adult will eat as many as 5,000 aphids in its lifetime.
- To attract a female, a male hedgehog will circle her while snorting.

WILDLIFE PROJECT: HOW TO MAKE A SOLITARY BEE HOTEL

By making a bee hotel, you can create a habitat for a wide range of solitary bee species. Red mason bees are active from April to June and use mud to line their nests. Leaf-cutter bees are active from June to August, and use pieces of rose, wisteria and birch leaves to line their nests – sometimes they use flower petals! Most solitary bees nest in cavities 8–12mm in diameter. The greater the range of hole sizes you can provide, the greater the diversity of solitary bees you will attract. Because gardeners remove dead wood from trees and shrubs, holes made by wood-boring beetles are less common than in the wild. This denies insects – such as some types of solitary bee – valuable nesting spots. In this project, holes drilled in wood mimic the natural process.

You'll need

Block of untreated wood
Drill and drill bits (1mm to 10mm in diameter)
Post to attach the block to
Sandpaper

Method

1 Place the wood on a firm surface and drill using a variety of drill bits to create differently sized holes.
2 Roll sandpaper into a tube and use to sand each hole, making sure there are no snags on the inside.
3 Thoroughly sand the block's surfaces so insects won't catch their bodies or wings on stray splinters.

> **TOP TIP**
>
> Once you've fixed the block on to a post, place it in a spot where it gets the morning sun – ideally facing south east.

Spotter's Guide to Wildflowers

Having developed an eye for garden plants, it's easy for us to miss some of the smaller, often more subtle – but equally lovely – wildflowers now jostling for the first pollinators. Naturally seeded into quiet corners around your garden or spotted on country walks, these are the cheerful heralds of spring, although some will still be with us through the summer too.

Wildflowers are far more variable than garden cultivars – in size, leaf density and colour. A deeper gene pool and broad adaptability to local geological conditions can result in lanky specimens or stunted runts. This means that plant sizes are difficult to be exact about, so in the descriptions below, low means prostrate, short is ankle height, medium is to the knees, and tall reaches up to your waist.

LOOK OUT FOR...

- **Cowslip (*Primula veris*)** – Short. Has soft, hairy stems with a nodding cluster of 10–30 deep-yellow flowers about 9–15mm in diameter. Leaves are broad and well-marked with pale veins. Dwells in dry, grassy banks, verges and meadows. Flowers April–May.
- **Herb Robert (*Geranium robertianum*)** – Short to medium height. Has light, airy foliage, the lower leaves with five subdivided lobes, often suffused with red. Hairy stems. Flowers are 14–18mm long, bright pink to white. Found in shade, hedgerows and woody banks. Flowers May–October.
- **Common bird's foot trefoil (*Lotus corniculatus*)** – This is a low to short, often sprawling plant, named for its seed pods that look like a bird's foot. Its small leaves form dense tufts. Its pert, custard-yellow flowers cluster in rings of 3–5. It likes rough, grassy places, banks and waysides. Flowers May–September.

- **Wood anemone (*Anemone nemorosa*)** – Low to short. Has jagged leaves with five deep lobes, and pinkish-red stalks and veins. Its 20–40mm wide flowers are crisp white, but often blushed with red under the petals. Found in hedgerows, woods and copses. Flowers March–April.
- **Bladder campion (*Silene vulgaris*)** – Medium to tall stems, topped by loose clusters of pure-white flowers (16–18mm), each erupting from an inflated bladder (the calyx). Its short, oval leaves are smooth and glossy. Found in dry grassy places, verges and field edges. Flowers May–August.
- **Red deadnettle (*Lamium purpureum*)** – Low to short. Its hairy stems bear triangular, crinkled leaves, the top ones often pink or purple. Its pink flowers, 10–18mm, are long-throated and hooded, with a lower petal poking out. Found in hedgerows and field edges. Flowers March–October.

SPANISH OR NATIVE BLUEBELLS?

How can you tell the difference? Look for the delicate, bending stem of the native, with bells dangling on one side, slimmer leaves and crucially, white pollen. Sturdier *H. hispanica* stands straighter, with light blue, pink or white flowers all round its upright stem and blue pollen.

Troubleshooting Guide

'I'm not sure there's any garden task that has a worse reputation than weeding, and thinking back to when I first started as a 16-year-old trainee, guess what job I hated the most? Yes, weeding! Today I'm more than happy spending a few hours weeding. There's something about this chore that I find really satisfying and therapeutic, and I love looking back at the ground I've worked on and seeing the difference I've made to my patch.

'Of course, weeding is not just about making your garden look good; it's also about improving the conditions for the plants you want to keep. If you do get overrun with weeds it won't only be tough for you, but also for your plants, as the weeds will soon start to steal water, nutrients and light, so it really is an important part of getting the best out of your garden.' – **Adam Frost**

TACKLING ANNUAL WEEDS

If you don't have time to give your area a full weed make sure weeds are never allowed to set seed. Snap off the flower heads, which helps in the long run. Hoeing beds on a regular basis, before weeds start to show, also really helps to save time further down the line. Come mid-season, when the garden has warmed up, grab an hour here and there in the week and hoe, but leave the weeds on the surface to dry then pick them up when you get spare time. In general, it's quicker to hoe an area on a dry day. Mulching the ground after weeding will also help save time in the long run as it suppresses weed growth. Try using a thick layer (around 8cm) of bark clippings, well-rotted manure or leafmould – this prevents weeds from growing, but also helps to retain moisture in the soil.

Did you know? Chickweed can set 1,300 seeds in six weeks.

HORSETAIL HELP

Field horsetail (*Equisetum arvense*) only rises to 1.5m but is very invasive. Deep rhizomes armed with tubers send up cone-bearing stems in spring, followed by green vegetative stems reaching peak growth in July. Management is your best bet. Remove spore-producing stems promptly and smother growth from spring to late summer to prevent it from fuelling the plant. Add mulches of well-rotted compost to raised beds and improve drainage. Line some raised beds with thick weed-suppressing membrane and fill with clean soil. Horsetail dislikes a raised pH, so try liming.

NORTH-FACING GARDENS

Although a north-facing site can be dark in winter, it may get more light than you think. The positions in the sky where the sun rises and sets change over the year, and between the spring and autumn equinox most north-facing walls get plenty of light – and in some cases direct sun – early in the morning and/or late in the evening. As long as you improve the soil with garden compost or manure, then you can choose from a vast range of shade-lovers – hellebores, heuchera and bergenias for low, evergreen foliage as well as flowers, and for summer colour try begonias and New Guinea busy Lizzies.

IMPROVING LAWNS

A few simple jobs will keep your lawn looking at its best. Rake out any thatch, using a spring tined rake or scarifier. This will help air circulate and encourages creeping grasses to thicken up. Collect any fallen leaves, which can make the grass turn yellow. Dig out any lawn weeds such as dandelions – dig down around them to get out the deep roots. Finally, adjust your mowing – aim to cut your grass to a height of about 2.5cm but don't remove more than a third of the height in one cut.

SAVING TULIPS

If you want to keep them, deadhead then water and feed (weekly with tomato food while the leaves are still green) to improve the display next year. Lift the bulbs from their container once the leaves are yellow, to store dry until autumn.

CROPS FOR SMALL SPACES

Try vertical growing! DIY or bought 'off the shelf', vertical planters can quickly green up any wall. If you don't have space for a veg bed, these are a great way of squeezing in edibles – position them near the back door for easy pickings. Chives, rosemary and strawberries are great for sunny spots, while kale, parsley and salad leaves suit shady areas. Remember to water regularly, as shallow containers dry out quickly.

CONTROLLING SLUGS

Regular slug hunts can be useful as most feeding is done after dusk, especially on damp evenings. In spring and autumn, collect and dispose of any slug eggs or rake over the soil at intervals so that birds can eat them. Drenching moist soil with a nematode slug control works brilliantly and controls soil-dwelling keeled slugs very well too. As well as being effective, these nematodes will not harm any other wildlife, pets or humans.

IMPROVING HOME COMPOST

Good compost needs a balance of ingredients including green waste (such as border trimmings, grass clipping and veg waste) and brown waste (such as brown paper, cardboard and dead herbaceous stems). Ideally add material in even layers. It should be turned regularly to mix the ingredients and add air.

JUNE

June is the month that the whole year aspires to. It is the culmination of spring and contains summer in all its billowing innocence before heat and time start to fray the garden at the edges. None of this is to do with specific weather. Hot June days are a joy, but the garden will still glow in a month that can be dogged by rain and unseasonal cold.

This is because June is dominated by green. Yes, there are lots of wonderful flowers – especially roses – but the intensity and freshness of green is what characterises our gardens throughout this month – from the lush grass, to the borders and new growth appearing on crops. The vegetable garden is burgeoning but – certainly in the first half of the month – is still slow to provide much of a harvest. This is why crops such as early broad beans or peas, which have been sown the previous autumn or at the very beginning of spring, are so welcome and taste so fresh and good.

But best of all are the long evenings, when we can stay outside gardening or just watching the swifts and swallows circle the sky, until after ten at night.

Weather Watch

We expect a lot of 'flaming' June. This is after all the month of the summer solstice, when the sun reaches its highest point in the sky and the peak of its warming powers. But hang on – what about airborne hats at Ascot, and 'play suspended' at Wimbledon? Rain-soaked garden fetes and Glastonbury mud baths? Blame the so-called European monsoon. In about three years out of four, the weather shakes itself out of its vague spring mood and robust depressions begin to march in from the Atlantic in a triumphant 'Return of the Westerlies'. The reason is most likely a knock-on effect of circulation changes elsewhere around the world.

WEATHER FACTS

- Average highest temperature in the UK: 18.5°C
- Average lowest temperature in the UK: 8°C
- Highest number of days of air frost: 0.6 in East Anglia
- Highest average number of hours of sun: 220.5 in the south east of England
- Highest rainfall: 100mm over 14.5 days in northern Scotland
- Lowest rainfall: 45.1mm over 8.2 days in London

WEATHER ALERT

Give plants support to protect against wind and downpours. Lush growth in summer borders is at risk of being flattened by 'monsoon' downpours, so ensure plants are supported to avoid damage. Tall structures covered in climbers such as sweet peas and runner beans may benefit from extra staking or even guy ropes.

 Star Plants

10 OF THE BEST

1. Peonies

Who doesn't love a full-skirted blowsy peony? Growing them can be a bit of a lottery, though. Get the weather absolutely right and you will have enough beautiful memories to carry you right through until the autumn. But if it rains at the wrong moment then those perfect blooms are more likely to be so waterlogged that they will look like bunched-up paper tissues. Yet, when it comes to flamboyant flowers few plants can beat the hardy, deciduous peony – they're worth the risk. Flowers May to June.

2. Mexican fleabane

Imagine if a crowd of slightly drunk pixies invaded your garden and charged around your borders carousing and giggling in high-pitched voices. They would look exactly like *Erigeron karvinskianus* flowers: an indomitable flurry of delicious white and pink daisies guaranteed to make you smile. Self-seeds happily in most situations – particularly good in gravel. Flowers May to October.

3. Foxgloves

The classic foxglove is a biennial, so expect leaves only in year one, flowers in year two, and then it sets seed before dying, leaving many offspring. The attractive spires add height to the backs of borders and they can be planted in shade, with flowers in shades of pink, purple white and red. Flowers May to July.

4. Roses

This is the month to celebrate roses. Try 'Burgundy Ice', a rose that is not only lovely in itself but also an exquisitely useful colour, with depth and

majesty like the velvet lining of a khan's best cloak. Its plum-coloured petals go well with the pastelly whites, blues and pinks that are all over our gardens at this time of year. A rich and generous flowerer that can cope with a little shade, although all roses need at least five hours' sun a day. It's a floribunda rose with glossy leaves and good disease resistance. Flowers June to October.

5. Turk's cap lilies

Turk's cap or martagon lilies are not the same as their big showy relatives – they are subtler and much more beautiful. They have the ability to turn a slightly shady corner into a seraglio of exotic delights. They also make very good cut flowers. Flowers June to July.

6. Alstroemeria

Alstroemeria have been part of our gardens for many generations. It is a plant that, if happy, is sometimes a little intrusive. The flowers appear in midsummer in colours that include red, orange, purple, pink and yellow, as well as softer shades of pink and white, providing waves of colour. They are also excellent in pots. Flowers June to July.

7. Astrantias

These pretty, clump-forming perennials produce pincushion-like flowers in colours from white through to a deep red and are a great choice for a shady border. Popular varieties include the deep red 'Claret' and the long-flowering *Astrantia major* 'Alba', which goes on until October. Flowers June to August.

8. Clematis

There is such a huge range of clematis to choose from including the long-flowering, white and purple 'Sieboldiana', which begins flowering in June, as well as the late-flowering clematis which flower from July. These take up little ground space but make a major impact when grown up fences and walls. Flowers June to October depending on variety.

9. *Primula vialii*

There are some plants that you simply cannot ignore – they jump out and ambush you as you saunter around the garden. This primula, Vial's primrose, is one of those plants. The combination of tomato red and mauve will grab even the most inattentive gardener. Best on the edge of a shady border (it will sulk if it dries out) or close to water. Flowers June to July.

10. Persicaria

Persicaria affinis, lesser knotweed, is a creeping, mat-forming perennial, with pale pink flower spikes and narrow green leaves. It's semi-evergreen and is an excellent groundcover plant in a mixed herbaceous border. It's also very popular with pollinators. Flowers June to August.

'If a plant can get two or three hours of light a day, then it will tolerate more shade than can be imagined.' – **Monty Don**

PLANTS FOR SHADE

- Erythroniums
- Ferns
- *Vinca minor*
- Solomon's seal
- *Alchemilla mollis*

WHY NOT TRY...

A new houseplant

Boston ferns are tropical plants with a real wow factor. The impressive leaves (or fronds) can grow to nearly a metre in length and will add a jungle or forest feel to any room. Hang them from the ceiling to let the leaves spill down or place them in a pot on the edge of a shelf to draw the eye. These plants are not only packed full of bold, bouncy foliage, they are also hard to kill, which makes them one of the most popular ferns grown in homes.

- **Position:** These do well in bathrooms and kitchens as they like high humidity, otherwise, mist it a few times a week.
- **Care:** Cut off dead fronds from the plant's base to allow new foliage to grow. Feed with half-strength balanced liquid fertiliser monthly from spring to early autumn.
- **Details:** Known to help clean the air, Boston ferns can absorb formaldehyde found in glues, paints, gas stoves and furniture.

 ## Jobs to Do this Month

Although June is a busy month, it's also a time to relax and enjoy the results of all your hard work. Whether you're deadheading roses on a summer's day or mowing the lawn, there are plenty of satisfying tasks to get on with. From continuing to sow seed, plant, feed and water plants and prune there's no shortage of things to put on the to-do list. Tender plants can be moved outside. It's a great time to plant a hanging basket and also a good month to enjoy long summer evenings surrounded by borders in full bloom.

KEY TASKS

- Deadhead roses.
- Keep mowing lawns.
- Sow seed of biennial flowers.
- Prune spring-flowering shrubs.
- Plant summer bedding.
- Harvest early potatoes.
- Prune evergreen shrubs.
- Plant out tender annuals.
- Look out for and tackle pests.
- Top up pond water.

- Neaten up lawn edges.
- Stake plants that have flopped.
- Hand pull weeds before they get big.
- Pinch out cordon sideshoots on tomatoes.
- Rub greenfly off stems and shoots.

GOT TEN MINUTES?

Deadhead spent blooms

When you snip off fading flowers, you are saving your plants from using energy-producing seeds and encouraging them to produce more flowers. It makes a big difference to garden tidiness too. For hybrid tea and floribunda roses, snip off individual flowers as they fade then after the last one cut back the stem to a healthy bud lower down. Don't deadhead roses with ornamental hips, flowers you want to produce seed or ornamental grasses.

GOT AN HOUR?

Plant up a hanging basket

This is a good month to create a hanging basket. The flower display is instant and it lasts until the first frosts of the autumn. Bedding plants like geraniums, fuchsias, petunias and begonias are ideal and you may find them on offer in the garden centres by now. Choose a mix of plants that will trail over the edge of the basket and strong uprights that will give the display some height. You can colour theme your flowers each year, too.

The shape of most baskets means that there is not much root room for the crowd of plants that share the space, so giving plenty of water and feed makes all the difference. Add extra fertiliser to the compost and use liquid feed too. Water daily through the summer and snip off the old flowers to encourage more. If the basket is way out of reach, rather than climb up to it to water, you can get a retractable pulley to lower it down for easy maintenance.

1 Use a large empty flowerpot to support the basket while you are working on it. Gather together the liner, compost, plants, plant food and watering can.

2 Line the basket with jute, leaving plenty above the rim, then an old compost bag to retain moisture. Fill with compost mixed with water-retaining gel.

3 Mix in slow-release fertiliser and begin planting the lower plants through holes in the liner. Next the high central plants, and finally, the plants around the edge.

4 Trim the liner around the edge and keep the basket on the pot while you place it on the ground and soak it thoroughly, allowing it to drain before hanging it in situ.

'I have always loved pelargoniums, partly for sentimental reasons, as they were the stock-in-trade of the nursery where I served my apprenticeship – and partly because, given decent growing conditions and a sunny spot, they flower for most of the year. Bedding, whether in beds, borders or pots, gives a boost to summer, reminding us that colour lifts our spirits.

'Pelargonium cuttings taken now will root rapidly, as long as they are given good drainage and kept in bright light. They are an easy and economical way to increase your stock of plants. A small propagator is helpful but not essential – even a windowsill can be used to raise new plants. Within a couple of months, they will be ready to pot on and overwinter for flowers next year.' – **Alan Titchmarsh**

GOT A MORNING?

Take pelargonium cuttings

1 Remove sturdy shoot tips from a healthy parent plant, making your cut with a sharp knife just above a leaf joint. Select only the strongest shoots.

2 Snap all but the top pair of leaves (one large and one smaller one) and any flower stalks.

3 Trim the cutting, making a cut immediately below a leaf joint.

4 Use a pencil to insert the cuttings around the edge of a pot of sandy compost. Good drainage is vital, as these cuttings tend to rot in soggy conditions.

5 Water them in and place the cuttings in a well-ventilated, plastic-domed propagator. Remove the dome when they start to grow and keep in good light at all times.

5 OTHER PLANTS TO TAKE CUTTINGS FROM

- **Begonias** – Long-flowering favourite, for sun or partial shade, in a range of cheery colours.
- **Fuchsias** – Fill your pots and borders with these dainty ballerina-like blooms.
- **Lobelia** – A great trailer, in shades of blue or white.
- **Petunias** – Grow plenty, for cascades of trumpet blooms.
- **Scaevola** – Fan-shaped flowers in blue, pink or white.

TOP TIP

Support tall perennials before they flop. Twiggy sticks poked into the soil hold up heavy blooms and foliage and look natural. Wires that link can be used to surround plants before they collapse. These solutions are best if placed before the plants need any support.

AND A FEW OTHER JOBS...

Trim clipped evergreens

Evergreen shrubs clipped into topiary shapes should be trimmed now to improve their shape. Otherwise they will grow leggy with sparse foliage. Use hand or powered shears to take this growth back to its neat lines. Put a sheet around the base to collect the clippings and don't cut back further than this year's growth. There will be another flourish of growth in the next few weeks, but the extra shoots from each cut you made will give a much finer surface to your topiary shape.

Plant out tender plants

Bed out the last of your tender plants, now that the frosts are over and night temperatures are consistently warm. The most temperature-sensitive flowers, such as lantana, lemon verbena and zinnias, as well as southern-hemisphere salvias and dahlias can now safely go out in pots or be planted in your borders. Give them a good soak when they are in position and deadhead them if they are in flower, because the burst of growth that follows this trimming back will help them acclimatise to their new summer home.

Water bedding plants

Water bedding plants early in the morning or in the evening to avoid sun scorch on their tender soft tissue. In every case, water them at their roots rather than spraying the foliage. And don't assume that after a rain shower they've had enough water – a long soak occasionally is better for plants rather than little and often. Plants in pots need more water than those in the ground, as their roots can't reach moisture deep in the soil.

Fill gaps with annuals

Sow hardy annuals such as calendula, cornflowers, cerinthe and nigella in any bare patches you won't be planting this summer. There's just time to benefit from the pretty flowers and you'll combat weeds, too.

 Time to Prune

With summer upon us, growth has been well under way for months on some plants, and cold snaps and frost may well have taken their toll on the new foliage, particularly of less hardy species. It's worth checking plants to see if there is any damage and cut back to healthy vigorous shoots or to a leaf joint.

The start of summer is your reminder to finish pruning spring-flowering shrubs. The last of these to flower are the likes of philadelphus and deutzia, which should be cut back as soon as the blooms start to fade. But it's not too late to cut back forsythia, flowering currant or weigela – just get it done soon.

It's also time to deadhead plants, cutting off faded blooms to stop them wasting energy on seed production (unless you want to save the seed) and, in the case of summer-flowering plants, to promote more flower buds. So keep those secateurs sharp and clean; there's always something that needs a prune.

PLANTS TO PRUNE

- Grapevine
- Flowering currant
- Rhododendrons
- Evergreens such as holly or *Viburnum tinus*
- Weigela (to reshape)
- Lilac
- Evergreen euonymus

In the Greenhouse

The greenhouse is consistently warm at this time of year, so there's plenty of opportunity to keep sowing, whether it's herbs or something more exotic like melons. As well as veg, it's a great time to sow biennials including foxgloves and wallflowers. It's important to remember to keep watering, especially tomatoes to prevent split fruits and blossom end rot. Use ventilation and shading to keep temperatures down on particularly hot days.

JOBS TO DO THIS MONTH

- Keep plants in good condition with some shading as they will scorch and overheat when the hot sun shines straight on them.
- Keep up with watering and feeding through the summer to guarantee healthy greenhouse plants. A general-purpose feed will suit most plants, but use a specialist formulation for orchids and cacti.
- If you still have room to plant more crops in the greenhouse border, go for those that will do well in higher temperatures. Aubergines,

chillies and sweet peppers will establish quickly, and produce their crop well into the autumn.

- On very hot days it's hard to keep greenhouses cool using vents and shading. So harness the cooling effect of water evaporation by allowing water to splash on the floor, to evaporate as the air heats up. Avoid doing this in the evening when the water will not evaporate, leading to fungal problems.
- Pinch out tomato sideshoots and tie in cordons securely.

DON'T FORGET TO...

- Mulch containers and soil, but only after watering plants thoroughly. It will suppress weeds and reduce water loss.
- Prune shrubs as they finish flowering, then water, feed and mulch.
- Keep up your weeding, by digging out perennials and hoeing off annuals.
- Train climbing and rambling roses.
- Cut back spring-flowering clematis – if they need it – now.
- Plant out potted lilies and cannas into enriched soil.

LAST CHANCE TO...

- Prune forsythia and weigela.

Fresh from the Garden

In June, there's the first flurry of significant harvests, the first sighting of colour, bursting from sweet pea blooms between spring-sown vegetables and perennial herbs – all announcing that summer is finally here. This month we can harvest the first handful of autumn-sown broad beans. That's followed by digging up the autumn-sown garlic, and then there is the popping open of young pea pods and enjoying the fresh sweet taste of what's inside. This month lettuce is ready to be cut, carrots and baby beetroot as well as the first potato harvest of Charlottes. There's also a bustle of bee activity around the flower buds of summer raspberry canes and strawberry plants – a sign that a fruity harvest is soon to come.

Remember to feed flowering plants (both vegetables and cut flowers) – a diluted feed of organic seaweed, alternated weekly with comfrey tea is a good option. It's also a time to keep on top of watering duties and get rid of any weeds that appear, including fat hen and chickweed.

WHAT TO SOW

- Salad
- Turnips
- Pak choi
- Beetroot
- Kohl rabi

WHAT TO PLANT

- Chicory
- Leek seedlings
- Outdoor tomatoes
- Tender veg

WHAT TO HARVEST

- Beetroot
- Chives
- Radishes
- Early potatoes
- Cherries

On the Veg Patch

- Sow fast growing crops like radishes and salad leaves to fill gaps in between slower growing crops.
- Trim strawberry runners to leave more energy for the main plant to develop a larger central crown.

- Tie in wall-trained fruit trees and remove shoots that are sticking out.
- Keep sowing French beans.
- Plant out aubergines, cucumbers, peppers and tomatoes into beds outdoors (after hardening off) or greenhouse beds.

CROP OF THE MONTH: GOOSEBERRIES

Did you know? – 'Goosegogs', as they're affectionately known, are among the hardiest and easiest of the soft fruits. Perfect for small spaces, they can be trained as fans or single-stemmed cordons against walls and fences, or as half-standards with room for herbs or salads beneath. If you have painful memories of prickles, there are plenty of spine-free varieties available.

Nutrition – Rich in vitamin C. Good levels of vitamin A and potassium as well as other minerals.

Storing – In the fridge, fruits keep for at least a week in shallow, covered containers. Freeze loosely set out on trays, then bag up when frozen.

Good cultivars – 'Invicta' AGM, the long-established green variety, for very heavy cropping and good disease resistance. 'Rokula' is early to mature, with heavy crops of sweet, dark red fruit, and good disease resistance. 'Xenia' has large red fruits, and is almost spine-free and resistant to mildew.

Plant – October–March

Harvest – June–July

HERB OF THE MONTH: BASIL

The warm, spicy, slightly minty flavour of basil complements many dishes, and aids digestion. An infusion of leaves can be drunk as a tea to ease a nervous stomach. Basil is ideal for growing in containers, as it is not invasive and looks attractive. It grows well with chervil, parsley, shiso and French tarragon, and improves the flavour of tomatoes, when grown together. Sow the seeds in early summer. Basil is a tender annual, and should not be exposed to frost. Start off seed in the greenhouse or on a kitchen windowsill.

GOT 15 MINUTES?

Harvest fruit

It's time for harvesting strawberries, red and white currants, gooseberries and blackcurrants. The first of the raspberries and blueberries will also be ripening now. Pick strawberries, gooseberries and raspberries individually. Hold the fruit gently to avoid bruising and discard any that are showing signs of rot. Pulling currants off the sprigs crushes the fruit so snip off the whole sprig then pick through them back in the kitchen ready for eating, cooking, bottling or freezing.

GOT HALF AN HOUR?

Plant out tomatoes

Tomato leaves will curl and distort in cold night temperatures, so plant now when warmer nights are likely. Harden off tomatoes before planting outdoors by leaving them outside during the daytime for a few days. Dig a hole, add well-rotted manure and chicken pellets, and mix in with the soil at the base before watering the hole thoroughly. Put the tomato in the hole and backfill with soil, firming well to leave the plant in a dip so the moisture from watering runs down to the roots. Support with a cane and tie. Water again and mulch with more compost.

GOT AN HOUR?

Grow lettuce leaves

We are all so used to popping into the supermarket to snatch a cellophane bag of ready-picked, ready-washed salad leaves, that it is all too easy to forget how simple – and satisfying – it is to grow them at home. They need little fuss, virtually no space and you know exactly what they have been through to get to your plate. Sowing now is ideal, as the weather is warm, the light intensity high and you will need nothing in the way of protected cultivation (no greenhouse or frame) to get them growing. It will only be a few weeks before you can harvest, and you can do that throughout the season on a cut-and-come-again basis. There are lots of varieties available, but a mixture is the best bet. Sow them every couple of weeks outdoors for a succession through the summer.

Fill a large terracotta pot with multipurpose compost. Terracotta keeps the compost cool and the moisture levels steady, which ensures even growth and a longer harvest time.

Sow the seeds thinly on the surface. They need to be far enough apart to allow for growth, but not so far apart that there are too few plants per pot, so place the seeds about 1cm apart.

Cover the seeds lightly with more compost. Use a fine sieve, or a small plastic pot with several holes in the base will do the same thing – cheap and effective!

Stand the pot in a tray of water until the surface becomes moist. This avoids displacing the seeds as can happen with a watering can. Stand the pot in a bright, sheltered spot.

ALAN TITCHMARSH'S TOP LETTUCE LEAVES

- **Lamb's lettuce** – Round, green leaves
- **'Lollo Rossa' lettuce** – Wine-purple leaves
- **Mesclun mix** – Sweet and tangy combinations
- **Rocket** – Finely cut, with a strong peppery taste
- **'Salad Bowl' lettuce** – Fresh, light-green leaves

OTHER JOBS

Sow French beans

Sow French bean seeds for a continual supply of tender beans right into autumn. Seeds can be sown straight into the soil outside, where you plan to crop them. If you have problems with mice eating the seeds, sow them into pots, then plant them out as seedlings. Sow two seeds at each station, 30cm apart, and bin the weakest seedlings. Support plants with low canes when they get heavy with beans. Pick them when they are young (and tastiest) and do this regularly so more will follow.

Protect fruit from birds

Make net covers to protect your ripening soft fruit. Use 20–25mm mesh to keep the birds out but still allow pollinating insects to get at the flowers. Put the netting on before the fruit starts to ripen – if you are too late, birds can strip a whole bush overnight. Make sure there are no gaps, because young thrushes and blackbirds are particularly good at finding holes and may get trapped.

Start training blackberries

Tie in the long shoots of blackberries and their hybrids onto a frame of horizontal wire supports. The vigorous new shoots emerging from the base will bear next year's fruit, so don't be tempted to cut them back. If you loosely tie them into your frame along with this year's fruiting canes, they'll be much easier to train in winter. Keep growth on the fruiting canes tied in and supported, as they will get heavy with fruit.

Water wisely

Some vegetable plants need more water than others. In general, you should soak plants at their base when they are newly planted and soak again in dry weather until they are well established. Spraying water around on the foliage is wasteful and not all that productive, whereas a good soaking at ground level creates strong drought-resistant plants with deep roots. The plants that you should target in dry weather in particular are tomatoes, cucumbers, squash, beans, celery and celeriac. It pays to get to know your crops – onions for example need water to grow and swell but just before harvest keeping them dry will improve their longevity.

Making compost

Firstly, get the right balance of carbon and nitrogen. As a rule of thumb, have as much dried stuff like stems, cardboard, straw or bracken as 'wet' material like kitchen waste or fresh lawn clippings – ideally more like two-to-one, dry to wet. Mix it up well, and if you can chop it up so much the better. When you spread compost on your garden you are not only adding material that visibly improves the soil appearance and structure but also adding millions – billions – of microorganisms that have both created the compost and made it their perfect home. Thus the soil, as it increases in organic material, becomes a rich loam that is packed with an unimaginable range and diversity of creatures all working together to support life – including your lovely plants.

'Get the health of your soil right and everything else follows and nothing – nothing at all – that you can do will contribute more to the health of your soil and the unseen but essential richness of life than adding your own garden compost.' – **Monty Don**

MONTY DON'S ADVICE ON SUPPORTING CORDON TOMATOES

Cordon tomatoes can easily grow to 1.8m tall and need support from the outset, let alone when they are laden with fruit later in summer. I often use bamboo canes, but if there is an overhead strut to tie to, supporting them with strong twine is the simplest and best way.

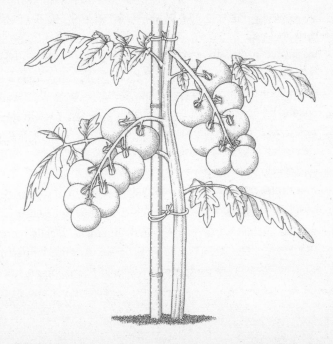

I tie the bottom of the twine to the stem just below the surface of the soil in the ground or pot, and the top to a bar or fitting on the roof of the greenhouse. As the roots grow, they anchor the twine firmly in place and as the cordon grows, I twist it round the string (as the twine is fixed it's much easier to do this than training the twine around the stem) and it is held flexibly but strongly upright. Then, when the time comes to clear the plants away in autumn, the twine goes on the compost heap along with the remains of the tomato plants.

COMPOSTING DOS AND DON'TS

- **Do** avoid too much of one single material.
- **Do** include a mixture of 'green' and 'brown' materials (see panel opposite).
- **Do** make sure the material stays damp. Have a watering can handy by a small heap and turn a hosepipe on a larger one in dry spells.
- **Do** trample it regularly to avoid air pockets that will allow portions of the heap to dry out and stop rotting.
- **Do** ensure your heap is held together in a strong container. In a small garden, consider using a compost tumbler.
- **Do** cover the top with a bit of old carpet or underlay to prevent moisture loss.
- **Do** invest in a shredder. It will make you a more assiduous and more effective composter!
- **Don't** expect lawn mowings on their own to turn into brown and crumbly compost.
- **Don't** add food waste for fear of attracting vermin – though uncooked vegetable peelings, crushed eggshells and unused salads such as lettuce leaves can be included.
- **Don't** compost thick-rooted perennial weeds.
- **Don't** add unshredded branches and thick twigs as they won't rot down.
- **Don't** use homemade compost in containers or for sowing seeds – it's not sufficiently sterile and its contents will be variable.

WHAT TO ADD TO YOUR COMPOST HEAP

To turn your garden waste into valuable compost, add a mixture of 'green' and 'brown' materials from the lists below.

GREEN

- grass clippings
- soft, leafy plants including annual weeds
- flowers
- fruit and vegetables
- uncooked kitchen waste such as vegetable peel, tea bags and coffee grounds
- bedding from vegetarian pets, such as hamsters and rabbits

BROWN

- prunings and hedge trimmings (ideally shredded)
- woodchip and sawdust
- wood ash
- dry leaves
- paper and cardboard (torn up or shredded)
- straw
- woody plant stems
- eggshells
- pine needles

TRY SOMETHING NEW...

Globe artichoke

This large architectural plant is impressive both in the garden and for its sweet and nutty flavour. It provides harvests this month. They can be grown from seed, but it's quicker to grow from pieces of tuber (root) from older plants and these can be planted from February to April.

How to: Choose a piece of tuber with a new shoot growing from it and trim back the old stem to just above the leaves. Plant in a sunny, sheltered spot that doesn't get waterlogged. Dig in well-rotted organic matter the autumn before planting. Place tuber, with the tip facing up and out of the soil, into a planting hole and backfill with soil. Firm in and water well. Cover the young plant with fleece until all risk of frost has passed. Mulch around the plant and, in summer, make sure the plant doesn't dry out. In the first year, your plant will be too young to produce a decent harvest, so cut off developing flower buds to help it gain vigour. As winter nears, cut back stems and wrap fleece round the plant. In the second year, cut off the smaller flowerheads, leaving one larger terminal bud in spring.

Rekha Mistry's Recipe of the Month

BROAD BEAN TORTILLA

You'll need

1 small onion
50g new potatoes (waxy), thinly sliced
50g shelled broad beans
1 clove of garlic (crushed)
½ tsp finely chopped mint
3 large free-range eggs (beaten)
2 tbsp olive oil
Salt and pepper to taste

Method

1 In a non-stick pan gently fry the onion until translucent.
2 Add garlic and sliced potatoes; mix gently together. Allow to cook with a lid on for 15 minutes over a gentle heat.
3 Add mint, season with salt and pepper and gently stir in the broad beans.
4 Pour in the beaten eggs and cook over a gentle heat until the middle of the tortilla is just set.
5 Place a plate over the top of the pan and carefully slide the omelette out, place the pan over the omelette and flip it back into the pan to cook the top side. Alternatively, you can finish cooking under the grill if your pan is oven proof.
6 Slice and serve, tapas-style, either hot or cold.

🦋 Wildlife Notes

With blackbirds feeding their babies on your lawn, blue and great tits scouring borders for caterpillars to feed nestlings, and manic buzzing in the borders, June is frantic with activity.

Leafcutter bee season is in full swing. There are three species likely to visit gardens: the patchwork leafcutter, *Megachile centuncularis*, the larger wood-carving leafcutter, *M. ligniseca*, and the Willughby's leaf-cutter bee, *M. willughbiella* – the male of which looks like he's wearing a big pair of gloves!

As solitary bees, they nest in bee hotels and other cavities – the Willughby may also use hanging baskets. They cut circular or oblong pieces from the leaves of plants such as birch, roses and wisteria to line their nest cells, into which they lay an egg on a parcel of pollen and nectar. Hang a bee hotel, with holes 10–12mm in diameter, in a sunny or partially shaded spot, 1–2m from the ground.

Young starlings are getting to grips with bird feeders. Trees are dotted with yellow-faced blue tits and hidden in hedges are sad-looking crea-tures with too much beak, too few feathers and a downwards smile. It's important not to disturb fledgling birds, particularly those sheltering beneath hedges. They have left the nest but can't yet fly. While they may look like they've been abandoned, they have nearly always been left deliberately by their parents. Trust birds to know how to raise their own young, and try to make sure you (and your cat) give the garden a wide berth during this time to avoid disturbing essential feeding trips.

LOOK OUT FOR...

- Male bumblebees, which are fluffier than the females and some have a yellow 'moustache', will be patrolling hedge bottoms for daughter queens.
- House martins. Listen out for their chattering calls as they fly catching insects to feed their young.
- Baby frogs and toads are leaving the pond now, so be careful when mowing the lawn.

- Baby birds. Many species will be fledging from nests this month. Look out for baby blackbirds on the lawn and blue tits that look like they've had a lemon wash.
- Swifts, which scream through the skies and nest in buildings. Erect a swift nest box for them to nest in next year.
- Garden tiger moths. This large moth has beautiful white upper wings with brown splodges, and orange underwings with blue-back markings.
- Leafcutter bees sealing cells of your bee hotel with sections of leaves.
- Red darter dragonflies laying eggs in your pond.
- Smooth newts sheltering in log piles or beneath stones.
- Peacock butterflies laying eggs on your nettles.

 ## Spotter's Guide to Pond Life

Adding a pond increases the wildlife value of a garden as it soon becomes a focus for creatures and will be investigated by skaters, beetles and dragonflies eager to colonise new sites. Make your pond in the open, in full sun, away from overhanging trees – falling leaves can affect the oxygen balance. To spot pond creatures use a small kitchen sieve and a plastic tray to tip water, weed and bugs into. When peeking in the pond don't let your shadow fall on the water, or try going out at night with a torch.

LOOK OUT FOR...

- **Water boatman** – Fast and agile, using oar-like rear legs to power itself through the water. Swims on its back and spends most of its time at the bottom but floats up, tail first, to replenish its air, which is stored in a bubble under its wing cases.
- **Great pond snail** – With a shell up to 40mm long, this is a veritable pond giant. Sometimes 'walks' upside down across the underside of the water meniscus. If disturbed, it emits an audible squelch as it expels air, then sinks down into the depths.
- **Pondskater** – Skittering quickly across the surface, it uses long middle and hind legs, which are tipped with fans of fine hairs to rest

on the surface. Short front legs catch and manipulate prey. Sometimes several gather around a victim.

- **Water hog-louse** – Really a freshwater woodlouse, doing the aquatic equivalent of snuffling in the silt and leaf litter at the bottom of the pool. It's adapted to living in water with low oxygen levels, so remarkably pollution tolerant.

- **Damselfly nymph** – Narrower and more delicate than the gothic-looking dragonfly larvae. It crawls through water weed, eating other invertebrates. Look for the paper-thin skins, left attached to stems, where the adults have emerged.

 Troubleshooting Guide

KEEPING PLANTS IN POTS HAPPY

Plants in pots have a limited amount of compost in which to sink their roots. This restricts both the food and water available to them. Dense foliage is a very effective umbrella, which means that you cannot rely on heavy rains to provide sufficient moisture for their survival. Regular watering (whenever the compost feels dry) and feeding once every week or so with dilute liquid feed during the growing season – from April to September – are necessary to keep them healthy. That said, overwatering (keeping the compost soggy) will result in plants that wilt, never to recover. This is exacerbated if the drainage holes at the base of the pot become blocked. In short, good drainage, sufficient water and regular feeding is the best recipe for healthy plants. And if possible, position your pot so it is sheltered from strong winds and yet still gets good light.

WHEN TO FEED

Plants will show signs of hunger if they're not fed and these signs should warn you they'll underperform. Many nutrient deficiencies result in leaf discolouration, with the patterns created indicating what's lacking.

As summer progresses and leaves age, the problems become more obvious, but you can prevent them from happening in the first place.

Plants in pots are most vulnerable as nutrients in the compost quickly get used up and are only replaced if you add fertiliser. Outdoors, the type of soil you have can result in nutrient deficiency, for example, acid soils make molybdenum unavailable to plants; likewise, iron with alkaline soils.

LILY BEETLES

With their shiny, pillar-box-red armour, lily beetles are among the most recognisable garden pests. You can find them now on lilies and fritillaries, eating notches out of leaf edges – they can even strip entire stems. While chemical solutions are available, you may need several treatments to gain control, and spraying plants while they're in flower will harm pollinators. Picking them off is effective if you're thorough, and do look out for the orange-red eggs and black larvae. A few lily types have shown resistance. Try *Lilium regale* or *L.* 'Defender Pink'.

It's worth paying attention to the creatures this month, while you can find them. Females are producing eggs this season, which hatch into voracious grubs. The beetles overwinter, well hidden in the soil. They're easy to forget until you display your magnificent lilies next year and notice ragged leaf edges and bright red invaders.

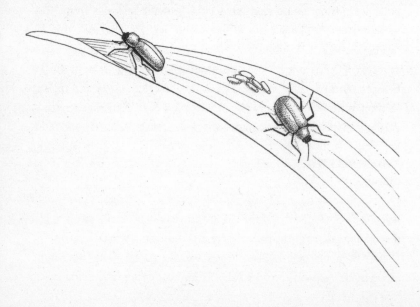

TOP TIP

Lily beetles fall to the ground exposing black undersides if you shake the plant. Place white paper underneath first.

POND ALGAE

A good way to reduce pond algae is to add oxygenating plants, such as hornwort (*Ceratophyllum demersum*), spiked water milfoil (*Myriophyllum spicatum*), willow moss (*Fontinalis antipyretica*) and water lilies, which shade the water and reduce algal growth.

CABBAGE CATERPILLARS

Butterflies and moths usually enhance our gardens, bringing colour and movement and adding wildlife diversity. But there's one group that causes havoc – the cabbage eaters. There are four main species, the larvae (caterpillars) of which feed on brassicas – Brussels sprouts, cabbages, cauliflowers, swede and turnip etc. Just eating a share of the vegetables we have grown wouldn't be too bad, if they didn't also leave frass (poo) behind and bore, unseen, into the tender hearts of our prize crops. They can be a pain.

Fortunately, control of these mini beasts is both straightforward and harmless. By covering crops with netting – do it now if you haven't already – you prevent access. You need to follow two rules: firstly, keep the netting away from the crop with hoops or poles, otherwise the insect will sit on top of the netting and lay eggs through it onto the crop; secondly, use netting with holes that are a maximum diameter of 7mm – surprisingly small. The eggs will then be laid elsewhere so that the caterpillars can feast on ornamental brassicas, such as honesty, nasturtiums or stocks.

BARE PATCHES IN THE LAWN

Rough up the soil surface to create a seedbed, sow grass seed (preferably the same mix as the original lawn), following the instructions on the box. Sow first one way, then at 90 degrees and not too thickly. Firm the soil surface over the seed with the head of a garden rake, water thoroughly using a watering can with a fine rose, then cover the area with horticultural fleece to maintain humidity and protect the grass from birds. Germination should be rapid at this time of year. Don't overwater or let the soil surface dry out. Alternatively, you could use a patch repair kit.

JULY

July is another season from June. Summer shifts and alters its angle. None of the freshness that June inherits from spring remains, but in its place there is a lushness and richness of colour, texture and sheer volume.

This is now high summer and the days are – very slowly – closing in. This affects the whole shape and tenor of the garden as the northern hemisphere flowers try desperately to seed from now on before winter arrives, but the plants from closer to the equator come into their own, relishing the warm nights that – hopefully – accompany the warm days. So dahlias, cannas and gingers all flourish and the tender annuals really kick in with a burst of colour that will carry through to autumn.

Weather Watch

The British summer has often been described as 'three fine days followed by a thunderstorm', and if that thunderstorm turns up on 15 July (St Swithun's Day) then, according to tradition, we can expect another 40 days of the same. However, supercomputers tend to be more reliable predictors than St Swithun. Still, thundery downpours often turn up this month, especially when warm, humid air drifts our way from the Mediterranean. The weather station at RHS Garden Wisley holds the UK record for the highest July temperature – a scorching 35.6°C on 19 July 2006. Remarkably, the previous night a ground frost had been recorded as far south as Shropshire! Exceptionally clear skies and very dry ground help create these huge temperature swings.

WEATHER FACTS

- Global warming? Since records began in 1913, the warmest July averages have all occurred in the last 40 years.
- Highest number of days of rain: northern Scotland, 15.7
- Days of air frost: zero
- Highest number of days of sun: 234, in the south east of England
- Average high temperature: 19.4°C
- Average low temperature: 10.9°C

WEATHER ALERT

Pot displays need plenty of water, especially on bright, breezy days. Glazed pots in full sun also get surprisingly hot, so turn them occasionally to avoid root damage. Terracotta pots, however, keep roots cool due to evaporation through the sides, although they dry out quicker, too.

 # Star Plants

10 OF THE BEST

1. Veronica

Borders, unlike most relationships, need a bit of spikiness to keep them exciting: all that soft, round foliage and those blissful, bowl-shaped roses need contrast. Something like a magnificent veronica, which introduces a bit of verticality (design language for spikes) to your garden. Try 'Marietta' which has the added advantage of being one of the bluest blues you could possibly imagine. Flowers June to August.

2. Phlox

Phlox are herbaceous border stalwarts – hardy and very easy to grow. They have sturdy stems, so even the taller varieties rarely need staking. They are excellent, low-maintenance plants for a herbaceous border or cottage garden that last for years. *Phlox paniculata* ('Blue Paradise') is the most commonly grown garden phlox. Everyone should try this variety –

the blue changes colour, becoming more purple in the day, then back to a violet blue as the light of long summer evenings fades to dusk. Really magical. Flowers July to August.

3. Nepeta

Catmint is one of those staple garden plants – the equivalent of pasta or rice in a store cupboard. Reliable, filling, colourful and pretty adaptable. You need to cut it to the ground as the first flowers fade to ensure that it bounces back. *Nepeta racemosa* 'Walker's Low' is easy to grow in any soil in sunshine and good on banks and border edges. Flowers June to September.

4. Perennial salvias

The most reliable and perhaps most intensely colourful of the salvias are the herbaceous perennials. Steady clump-formers, they have vertical stems that are packed with two-lipped flowers for many weeks, or even months, and add a welcome concentration of colour while retaining an elegant look. Try *Salvia nemorosa* 'Ostfriesland', an award-winning variety with purple stems that carry dense spikes of violet flowers for many weeks. Flowers July to September.

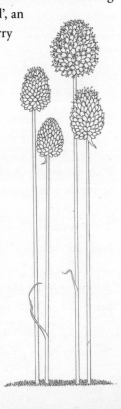

5. Lavender

Easy to grow, lavender thrives in sunny sites and free-draining soil, but annual pruning is essential to prevent the plants becoming straggly. Prune hard every spring, or in August. For something different, try 'Arctic Snow' – spikes of pure white flowers sit closely above fresh, green foliage on bushy and well-branched plants. Flowers July to September.

6. *Allium sphaerocephalon*

In May, we were awash with alliums: white ones, purple ones and pink ones. Some with flowers like sputniks, others with blooms as big as a bundled lemur. They are a big part of our early-summer gardens but, by July, they have faded to handsome but less colourful seed heads. Except they haven't all gone over, as in July we have the pleasure of the drumstick allium. Flowers July to August.

7. Achilleas

The bold, flat heads of achilleas are really useful at this time of year. These garden-bred varieties are descended from the yarrow, a native grassland plant, so they go well with ornamental grasses. They give a soft meadow effect, with a bit of zingy colour as an extra benefit. Try the orange 'Terracotta'; it's bright colour fades to yellow as the summer progresses. It's good on poor, free-draining soil. Flowers June to September.

8. Cosmos

Cosmos are sun-loving plants with a long flowering season. The elegant, simple single-flowered types are always popular, with traditional colours, antique shades and peachy bicolours, mixed with prolific flowering. And they're easy to grow. 'Sonata White' is a good variety for pots, short and prolific, with bright white petals and a golden centre. Flowers July to October.

9. Japanese anemones

Japanese anemones like 'Pretty Lady Emily' add height and colour to a late-summer border. This cultivar is one of a series of Pretty Lady cultivars developed in Japan and they are more compact than other varieties. It's a good semi-evergreen plant for pots and borders. Flowers July to September, although some cultivars flower later.

10. Monardas

Look out for the distinctive flowerheads of bergamot such as 'Gardenview Scarlet' or 'Scorpion'. Each has a large number of curving tubular flowers growing out from a central point, creating a shaggy dome of petals. It's great for borders as it has a long flowering season, blooming almost continuously if deadheaded. The foliage is aromatic, and leaves are sometimes picked for pot-pourri. Flowers June to September.

> *'In nature, Mother Earth makes all the decisions. In our gardens, we get to choose which plants to put together.'* – **Carol Klein**

| **WHY NOT TRY...**

Carol Klein's plant combinations for summer

- **Dahlia and canna** – A simple contrast of colour and form makes a dramatic statement. *Dahlia* 'David Howard', with its almost spherical, apricot flowers, complements the bold, paddle-shaped leaves of the *Canna* 'Wyoming'. Later, it too will produce similarly coloured flowers, but of a totally different form.
- **Astilbe, aster and monkshood** – The frothy white fronds of *Astilbe* 'Professor van der Wielen' keep company with aster (*Symphyotrichum* 'Little Carlow') and spires of white monkshood (*Aconitum* 'Ivorine') are perfect for adding extra interest.
- **Astrantia, geranium and bears breeches** – Astrantia, a soft grey *Geranium pratense* seedling, and an invading bears breeches (acanthus) make an attractive combination, reflecting each other's colours while displaying different forms.

PLANTS FOR EVENING SCENT

- Honeysuckle
- Common jasmine
- Night phlox
- Tobacco plant
- Sweet rocket

Jobs to Do this Month

In high summer, it's all about keeping the show going – many flowers and veg are at their peak, so there's plenty of watering, deadheading and feeding to do to ensure displays look their best. Sweet peas, roses, summer bedding should all be looking vibrant and colourful, there are long evenings to enjoy but this is also a good time to look ahead – both short term, planning for any holidays and how to keep your plants healthy while you're away, as well as for the next season. Why not plant some autumn-flowering bulbs or take some photos of the garden as it is now so you can review your planting in the winter?

KEY TASKS

- Move pots before going away.
- Water your compost bin.
- Give shrubs a feed.
- Deadhead dahlias.
- Pick sweet peas.
- Plant autumn crocus bulbs.
- Water patio containers.
- Cut back early summer perennials.

GOT HALF AN HOUR?

Cut flowers for vases

Avoid the heat of the day for picking your blooms and cut them with as long a stem as possible. Strip the lower leaves, then plunge straight into a deep bucket of cold water for a few hours before re-cutting and arranging. Picked flowers can last for days if they are conditioned properly before you put them in a vase. Seal sappy stems with a quick dip in boiling water. Woody stems do best with a vertical split up through the base and hollow stems fill with water more efficiently if the water is a little warm. Gladioli, lilies, pinks, rudbeckia, sweet peas and zinnias give long-lasting blooms for several days.

GOT AN HOUR?

Make more bearded irises

Divide bearded irises once they have finished flowering to invigorate the display for next year. Cut back the old flower stems, then lift them and use a knife to divide the fleshy rhizome into smaller sections – make sure there are sets of leaves on each section. Plant them back or pot them up, keeping the rhizome exposed on the surface to stop it rotting. Trim the leaves, too, and more will soon develop to strengthen the plant ready for next year's blooms.

GOT A MORNING?

Take softwood cuttings

Rosemary is a great plant to take cuttings from. They root very easily, so you will have new plants in no time.

1 Select a healthy rosemary shoot, with no flowers, about 10–15cm long and cut from the plant using a sharp pair of secateurs or scissors. Place the cuttings in a plastic bag to stop them drying out before you plant them.

2 Pull off most of the lower leaves so that there is a length of clear stem. This stops the cutting loosing too much moisture. Using a sharp knife, cut off the base of the stem just below a leaf joint.

3 Fill a 10cm pot with compost. Using a thin cane, make a hole in the soil at the edge of the pot roughly the same depth as the clear stem of the cutting. Put the cutting into the hole down to where the first leaves start and firm in.

4 Water the cuttings thoroughly using a rose attachment so soil isn't washed away. Add a label with the plant's name and place in a cold frame or greenhouse, or in a propagator, or covered with a clear plastic bag indoors.

'I can remember the first time I took a cutting with my nan. It was from a coleus, and when it rooted after a couple of weeks I felt like it was something magical! I still feel like that now, and there aren't many things more satisfying than creating new plants. People are often afraid to take cuttings because they think it seems complicated, but it's easy to do and can be done with so many plants. Now is a great time to give it a go as the plants are actively growing, so your cuttings will root easily.' – **Adam Frost**

---

5 MORE PLANTS TO TAKE CUTTINGS FROM

- Penstemon
- Salvia
- Fuchsia
- Pelargonium
- Clematis

AND A FEW OTHER JOBS...

Collect and save seeds

Wait for a clear, dry day then collect some seeds of flowering plants like lychnis, aquilegia, foxgloves, knautia and centranthus to sow later in the season or next spring. Many early-flowering perennials will have dry seed capsules by now and these are the easiest to harvest and store. Snip the stem below the seed heads and place them upside down into a paper bag or envelope. The seeds will release into the bottom; the capsules can be discarded. The seed must be stored dry. Label the envelopes and keep safe in a tin.

Tend pond plants

A pond can have a mixture of plants under the water, floating across the surface and emerging from the edges. These will grow quickly in summer's warm water and some can be invasive, so will need to be tended just as you would with plants in borders. Deadhead flowering pond plants to encourage them to keep flowering for longer. You need to cut off the blooms before they fall off to rot in the water. Remove unsightly dying leaves regularly before they start to rot and deplete the water of oxygen at a crucial time now the pond is full of creatures.

Tie in and support climbing roses

There is always a point in July when the climbing roses become heavy with flower and the new growth, which produces the flower buds, is not yet ripe enough to support them properly so they flop, and if there is a storm, can easily be damaged. I always go round them and loosely tie them all in to the supporting wires with twine. I do not prune or try and shape the plant, but simply give it the support it needs before the shoots are woody enough to bear the weight of their own flowers. By autumn this new growth will have hardened off and I can make careful decisions about what to cut and what to leave behind for next season's framework – which in turn will produce its soft new growth.' – **Monty Don**

Plan your watering

Prioritise the plants that need water most as it's a precious resource. Containers should be watered almost every day – even when it rains, the water will drop off the leaves with little reaching the roots. The most efficient way to water is to take off the rose on the watering can and direct all the water onto the plant's root zone. In borders, target plants that are newly planted and soak the ground around them every few days. If a scorching-hot day is forecast, do it early before the water can evaporate.

Deadhead bedding

Pinch out flowers that have started to fade, by hand. Spend 10 minutes every day enjoying this bit of light gardening – it's a great time to reflect and you'll really notice the difference. Bedding plants normally flower right through the summer, but they'll do far better and look much tidier if you take time to deadhead them, otherwise some plants will just stop flowering and go to seed.

TOP TIP

Garden centres and nurseries are usually selling off the last of the summer flowers this month at bargain prices. Choose hardy perennials for flowers next year then trim them down low and soak thoroughly. Repot or plant out and keep watered while they get established.

Plant autumn bulbs

Buy and plant autumn-flowering bulbs into the borders now. Generally, dry bulbs are supplied in the spring but many nurseries and garden centres also pot grow this group so you can add them to the borders in the green, with confidence that neighbouring plants won't swamp the young shoots as they emerge. Nerines and crinum need plenty of space in a well-drained sunny position with bulbs slightly exposed at the surface. Colchicum, cyclamen and the lovely autumn-flowering snowdrop can cope with some shade and do best in humus-rich soil.

DON'T FORGET TO...

- Pick sweet peas every 10 days or so to encourage more blooms.
- Dig up elephant garlic and leave it to dry until the tops have withered.
- Look for the conical spent flowers of dahlias and deadhead them.
- Tie in growth on roses to support large flowers.
- Get any remaining autumn-flowering bulbs into the ground.
- Keep the areas around newly planted trees and shrubs free of weeds and well watered.

LAST CHANCE TO...

- Get any remaining autumn bulbs into the ground.

Time to Prune

At the height of summer, the fresh greens of spring and early summer have been replaced by subtly darker shades, which indicate that foliage has toughened in response to less available water, longer day length and the sun being directly overhead at midday. In fact, in very dry weather, the young growth of woody plants may be wilting, so use your secateurs to remove the top third of these vulnerable shoots to reduce the area of foliage that can lose moisture. This simple technique is also good for fruit trees and bushes, where the current season's shoots can be cut back to just above the lowest two buds. This will channel the plant's energy into forming more flowering buds for the following season, swelling the developing fruit and exposing the fruit to the sun for ripening. Summer pruning is also essential for vigorous climbers, and spring-flowering ornamental trees and shrubs to maximise their flowering potential.

PLANTS TO PRUNE

- Deutzia
- Rhododendrons
- Weigela
- Rambling roses – after flowering
- Apples and pears
- Summer-flowering shrubs – after flowering

TIDYING UP

Spring and early summer-flowering herbaceous perennials can start to look tatty. Cut back the flowered stems down to the base and carefully nip off any yellowing or dead foliage in the clump. Allow the rest of the foliage to feed the plant so that it can bulk up for next spring. In some cases, you may get a second flush of flowers in late summer.

In the Greenhouse

It is hopefully warm and sunny this month, which means it's important to keep an eye on greenhouse temperatures. Fungal disease thrives in humid air so it is worth ventilating the greenhouse to get some air exchange, even though it will increase the plants' need for water. Ventilation also reduces the extreme differences in day and night temperatures that add to plant stress. Automatic roof vents and side louvres are the best combination as, when they're open, air is drawn gently up through the house. On the hottest days, keep the doors open too, use shading and damp down the paths or soil surface with water as this will lower the overall temperature. Plants are growing rapidly now, especially in the heat of the greenhouse. Fertiliser in most composts will feed a plant for around six weeks so you will need to feed anything potted up for longer than that. There's no shortage of jobs to do in the greenhouse this month…

JOBS TO DO THIS MONTH

- Give extra care to any plants you transfer outside. They can suddenly wilt when exposed to sun and wind, so water well.
- Blossom end rot causes unripened circular patches at the base of tomatoes or undeveloped ends of fruit on cucumbers, courgettes or squash. To prevent this, apply liquid feed and never allow the compost or soil to dry out completely. The rot indicates a calcium deficiency, not caused by a lack of calcium in the compost, but by water not flowing well and delivering nutrients right to the plant's extremities.
- Help tomatoes to ripen by removing lower leaves from plants. This allows plenty of light to reach the fruits, which tomatoes need to colour up. Lower foliage is no longer important to the plant as the most productive leaves are at the top, so cut back the leaves, clearing the stem from the ground to the fruit. Continue to do that up the stem as the higher trusses ripen. Reduce the watering a bit too, to improve the flavour of the fruits now there are fewer leaves.

- Biennials sown last month will now be ready for potting on. Given nutritious potting compost and a bit of space, they will grow into strong young plants ready to plant outside in autumn. Forget-me-nots, foxgloves and wallflowers all need this treatment to ensure a good flowering display next spring.
- Water nematodes onto the compost if you spot vine-weevil leaf notches.

Fresh from the Garden

Summer is now in full swing – in just a few months, a veg patch can go from bare soil to a hive of activity, with worms burrowing in the warm soil, bees buzzing on blossom, and hoverflies and ladybirds munching aphids on runner bean and dahlia flowers. The veg harvests roll through July with increasing profusion: the peas and beans, salad, beetroot, carrots, artichokes, summer cabbage, cauliflower and kale – not to mention the strawberries, raspberries, red and blackcurrants and gooseberries – are all at their best. As well as keeping on top of all these harvests, remember the weeds – the annual weeds grow fast, so weeding the soft, warm soil between rows of veg is essential.

WHAT TO SOW

- Chicory
- Spring cabbages
- French beans
- Spring cabbages
- Turnips

WHAT TO PLANT

- Leeks
- Brassicas

WHAT TO HARVEST

- Beetroot

- Carrots
- Courgettes
- Garlic
- Tomatoes
- Onions
- Peas
- Salad leaves
- Soft fruit
- Spinach

'What draws me to the plot? It's the early-morning bird chorus and warm, sunny days, filled with the aromatic scent of sweet-pea flowers. Everything's in bloom and putting on a marvellous show, from vegetables to soft fruit and cut flowers such as cornflowers and zinnias.' – **Rekha Mistry**

KEY TASKS

- Look after runner beans.
- Harvest garlic.
- Thin out apples.
- Water veg in pots regularly.
- Keep picking greenhouse crops like tomatoes and chillies.
- Earth up the stems of brassicas to keep them stable.
- Hoe between crops once a week to prevent weeds flowering.

CROP OF THE MONTH: RADISHES

Did you know? – Radish seed pods are as delicious as the roots. Leave plants to bolt, flower and set seed, then pick the pods green and eat raw for a crunchy, radishy hit. Any radish will do, but 'rat's tail' varieties have particularly plentiful pods.

Nutrition – The roots are packed with antioxidants, as well as containing high levels of vitamin C and folic acid.

Storing – Radishes are at their crunchiest eaten minutes after harvesting, so pull them as you need them. Otherwise, they'll stay in good condition in the fridge for about a week.

Good cultivars – 'Sparkler' (cherry-sized roots); 'French Breakfast' (elegant and elongated); 'Black Spanish' (black-skinned winter variety)

Sow – March–September

Harvest – April–October

HERB OF THE MONTH: MINT

The delicious, clear, spearmint flavour of classic garden mint is wonderful with strawberries, yoghurt, marinades and dressings, or as the main ingredient of mint sauce and jelly. Mint is also a renowned digestive. Being a root runner, it is best grown in a container on its own. If combining with other plants, sink it in its pot – this will stop it taking over the display. It combines well with calendula, chamomile, chives and sage. Take stem tip cuttings in spring or early summer, or root cuttings in autumn. Like all vigorous herbaceous perennials, mint benefits from being divided every one or two years.

GOT 15 MINUTES?

Harvest crops

The midsummer harvest can be overwhelming but it's vital to keep picking vegetables to stop them going to seed. Leafy salads, herbs and spinach should be cut, then watered to keep them in growth – if allowed to dry out, they will bolt and make the remaining leaves unusable. Dig up potatoes as you need them. Pick tomatoes and courgettes so they bear new fruit. Harvest beans and peas daily – even hardened and inedible pods should be snipped off and discarded. This will stimulate new flowers and more fresh pods for later on. Use up gluts by making chutney, and sauces and soups for freezing.

GOT HALF AN HOUR?

Thin apples

Reduce the number of small fruits per cluster down to one or two to improve the size and quality of the apples, and keep the tree in good health for future fruiting. Leave the best-formed fruit in each cluster. From June onwards you may have noticed a few immature fruits starting to drop from plants as part of a natural process. However, you may also need to do some manual thinning.

GOT AN HOUR?

Train climbing veg

Use a frame support to train climbing crops like beans, melons and summer squash. Tie in where necessary and pinch out any shoots that cannot be supported. Pinch out the sideshoots of vine tomatoes and tie in upper growth every few days to keep them as tall, single stems rather than sprawling bushes. After your gooseberries and redcurrants have been picked, trim their side shoots to a short spur. Retie to the support where necessary and be ready to train the vigorous new growth.

OTHER JOBS

Harvest garlic

Dig up your garlic now that the foliage has yellowed and begun to collapse. It must be harvested at once and should be carefully stored to keep it plump and fresh for as long as possible. Keep the stems intact, and dry the papery bulbs for a day or two, then plait them (if you wish). Hang them somewhere airy and dry where you can snip them off for cooking. You can also peel the cloves and freeze them or preserve them in oil.

TOP TIP

Look out for bolted crops – these are crops that have flowered early. At this point, succulent foliage becomes coarse and bitter. Harvest the leaves to eat then pull out the plants and sow another crop.

Pinch out cordon tomatoes

Regularly examine the axils between the stem and leaves on your cordon tomato plants to check for sideshoots. Pinch them out as soon as they appear to stop them turning your plant into an unruly mass of stems. It's important to do this when they are small, as larger sideshoots can tear the stem when you eventually remove them. Keep training the plant straight up, with one single stem, supported by a cane or string. It will keep producing sideshoots, so be vigilant and act quickly.

Sow kohlrabi

Sow shallow drills of kohlrabi now. Flea beetle often devastates earlier sowings, so the quality will be better now they've gone. Kohlrabi is fast from sowing to harvest, so keep a perfect supply going by sowing just a few every couple of weeks. Keep them well watered. They're best eaten young when the bulbous stems taste like a tender cabbage heart.

Feed crops

Sprinkle a few handfuls of granular fertiliser on the vegetable patch to stop your veg running out of steam, especially if you have big croppers like courgettes, squash and tomatoes, which are hungry feeders. Brassicas, celeriac and leeks will keep growing for autumn and winter cropping, so feed those, too. Fork or hoe lightly around the plants to weed and loosen the soil, then apply the fertiliser before soaking, if there is no rain forecast.

TRY SOMETHING NEW...

Malabar spinach

Malabar spinach is a tender perennial vine, and its thick, fleshy leaves make a great alternative to spinach or chard. Buy seeds online and sow in March–May to harvest in July–September. As it's a tropical plant, it thrives undercover or in greenhouses, but it can also be grown outdoors in a warm spot. Harvest the leaves when they are young to add to salads and cook the mature leaves. At every leaf axil, a flower stalk is produced, covered with pink flowers containing black seeds, which you can use to sow the next crop – or take cuttings.

 # Rekha Mistry's Recipe of the Month

MALABAR SPINACH SAUCE

You'll need

2 tbsp sunflower oil

1 medium aubergine, peeled and diced

2 plum tomatoes, diced

2 tbsp each of ginger and garlic paste

2–3 fresh chillies, roughly chopped

1 tsp each of mustard, cumin and coriander powder

½ tsp chilli powder (optional)

500g Malabar spinach, roughly chopped (if unavailable, use spinach or chard greens)

1 tsp tomato paste

Salt to taste

Method

1 Heat the oil in a pan and sauté the aubergine. Once seared all over, add a splash of water, then cover with a lid.
2 Add the tomatoes and all the spices, and let it sweat down on a low-medium heat, stirring occasionally. Add the spinach, tomato paste and a little water, and cook for 15 minutes.
3 Allow to cool, then blend carefully until smooth.
4 Return to the pan. Add water if the sauce is too thick or cook for longer if it is too thin.
5 Serve with spice-stuffed potatoes, grilled halloumi, fish, chicken or sizzling prawns.

Wildlife Notes

July is the perfect time to sit outside and enjoy the garden in the evening. On mild, dry nights, hide in a quiet corner to keep an eye out for nocturnal creatures springing into action. The most active part of your garden will be the pond. If you have one, look out for foxes and hedgehogs coming to drink at the water's edge. Pipistrelle bats can congregate at ponds, flitting above the water to catch insects such as mosquitoes. If your pond is large enough, you may attract Daubenton's bats, which scoop insects from the surface with their hairy feet. Shine a torch into the water to see backswimmers (*Notonecta* species) hunting for invertebrates, and spot amphibians such as newts.

Most birds have now fledged their nests. In and around our homes, swifts, swallows and house martins continue to feed their young. House sparrows may be on a second, or even third brood, while finches still nest in hedges. On the whole, things are quieter, but there's new wildlife emerging. It's the start of garden butterfly season. The second genera-tion of small tortoiseshells, red admirals, peacocks and commas will be landing on buddleias. They need sunny days but a bit of rain – a good, old-fashioned British summer.

LOOK OUT FOR...

- Dragonflies and damselflies. The males will be patrolling gardens now, while the females (sometimes still attached to a male in what's known as a 'mating wheel') may lay eggs in your pond.
- Grasshoppers. If you're lucky enough to have long grass in the garden then listen out for their gentle chirruping. They 'sing' by rubbing their legs against their wings.
- Bats might be spotted flying around at night, catching insects. Look for them around the tops of deciduous trees and above ponds.
- Butterflies such as the peacock, red admiral and small tortoiseshell. The second generation of adults is hatching now.
- Baby frogs, toads and newts leaving the pond. Make sure there's plenty of long grass and low-growing plants for shelter.
- Swifts, swallows and house martins, which are still nesting. House martins may continue nesting until September.
- Common wasp workers snatching caterpillars from your brassicas.
- A nest of field mice or voles in compost bins or sheds.
- Bats catching insects in the treetops.

- Hornets hunting bees.
- Bottle-green rose chafers flying clumsily around your roses.

🔭 Spotter's Guide to Solitary Bees

Unlike honeybees or bumbles, most bee species are 'solitary' and do not live in organised colonies. Instead, each female bee builds her own individual burrow nest, in soil, sand or in a pre-existing tunnel in dead wood, loose mortar or hollow stem. Some smaller species nest in empty snail shells. Although bumblebees are well recognised, solitary bees are also vital pollinators.

Solitary bees often nest near each other, digging into a sunny bank, or using old beetle burrows in a dead tree trunk, giving rise to the charming idea of 'bee villages'. Digging, nesting, foraging and laying up of a nectar/pollen cake for the offspring are done by females, but the smaller, often darker, males are usually about, too, hawking about the nest sites awaiting opportunities to mate. Some solitary bees are harder to spot, and some species need our help. Leaving aside worries about neonicotinoid pesticides, many bees are vanishing because of changing land management.

The loss of flowery hay meadows, in favour of intensively farmed silage fields, has reduced foraging opportunities. Urban encroachment also degrades vulnerable habitats, especially in coastal areas where land unsuitable for agriculture is seen as 'waste' ripe for building development. With over 250 bee species in Britain – 225 of them solitary bees – a worrying number of them are diminishing.

5 COMMON SOLITARY BEES TO LOOK OUT FOR...

- **Red mason bee** – Broad, square-tailed, ruddy brown; female has two distinctive horns on face. Nests in old walls or in rotten timber. Readily takes to sections of (cut) bamboo in bee 'hotels'.
- **Leafcutter bee** – Stout, broad bees; several species. Abdomen has coat of pale hairs underneath to collect pollen. Cuts segments of leaf to line nest in soil; flies in a zigzag.

- **White-faced bee** – Small, slim, shining, almost hairless, males have round white faces. Females are virtually all black but with white-spotted faces. Nests in dead, hollow stems.
- **Cuckoo bee** – Many species, strikingly marked with black, yellow and red bars. Almost hairless and often mistaken for a wasp. Sneaks in to lay its eggs in the nests of other solitary bees.
- **Bronze sweat bee** – Small, black with greenish bronze thorax. Named after their habit of visiting moist arms to drink sweat, these fast, active bees nest in large aggregations in bare soil.

Troubleshooting Guide

'Try as we may, there are always things that don't go according to plan – areas of the garden that aren't performing as we had hoped. But with a little ingenuity and forward planning, many of these irritations can be solved. Crack the problem and your summer will be more enjoyable as a result.' – **Alan Titchmarsh**

BIRDS EATING FRUIT

There is only one way to make sure that you – and not the birds – eat your raspberries and strawberries, and that is to cover them with netting. Not loose netting in which the birds can become entangled, but a proper fruit cage (tall for raspberries, lower for strawberries). It really is as simple as that, and the netting can be removed and stored after cropping so that your garden doesn't look as though it has its own tennis court.

PLANTS HAVE STOPPED FLOWERING

Many plants – such as sweet peas – will stop flowering if they are allowed to run to seed. Keep picking flowers to bring indoors, and deadhead regularly to keep plants flowering right through the summer and into autumn. Not only is this technique useful for prolonging flowering, but the garden will also look much better for being free of faded flowers. Verbena flowers for a long time even without deadheading, but you may find it spreads its seed a little too freely. So, remove faded flowers to keep it looking neat and stop it popping up where it's not wanted.

INSECT ATTACKS

If you are organic, you will need to look at ways of limiting attack – including growing a wide range of plants, mixing them up and avoiding plants that are martyrs to greenfly or whitefly. Squirt off pests with a powerful jet from a hose, and encourage birds (many of which eat insects) to come into your garden. Ladybirds and other beneficial insects can be encouraged by planting pollen-rich flowers. Wasps do their bit in controlling plant pests, so don't think of them entirely as villains (see over the page). If you choose to spray with an insecticide, bear in mind that you are probably also killing beneficial species. Learn when to turn a blind eye – plants that are well grown, well fed and watered, are more capable of shrugging off attack than feeble ones that are struggling.

WASPS

They are the gardener's friend – eating insect pests and caterpillars – but they also eat fruit and make outdoor dining a pain in late summer. Trap them in narrow-necked jars of jam and water, or try hanging up those paper imitation wasps' nests which are reputedly effective.

MILDEW AND BLACKSPOT ON ROSES

These fungus diseases often strike when the plants are under stress – usually on account of a lack of food and water. Climbers against house walls are especially prone to attack. Make sure that food and water are in good supply, and mulch the ground around the plants with well-rotted manure, compost or chipped bark to retain moisture in dry spells. If your varieties are prone to these diseases, swap them for leathery-leaved, disease-resistant varieties, which are less likely to be attacked. That way, you and your plants will be happier.

MOLES

There are all manner of tricks you can try to discourage moles: mix powdered mothballs with the molehills, push children's windmills into them so that the vibration deters the mole, or sink empty wine bottles – neck upwards – in the hills so that the wind whistling across the top sends him packing. There are special 'mole movers' – vibrating metal spikes that can be pushed into the ground and that work reasonably well on clay soil but are not particularly effective on sandy earth. In extreme circumstances, some people may resort to calling in a mole catcher.

AUGUST

Even if the weather in a British August is overcast and stormy, there is always a sense of summer, holiday sun being the rightful and natural order of things. The nights are generally becoming cooler so dew refreshes the hottest days and the hardworking gardener can sleep easier. What changes more than anything is the quality of light as the sun slips lower in the late-summer sky. This means it's the richer colours that dominate our borders: purples, burgundies, bronzes, golds and oranges shine out, especially in the low-cast evening light. But there is also a slight tiredness in the foliage still robustly attached to trees and shrubs. The answer is to make sure that all hedges are cut, and undue shagginess pruned and tidied. This will spruce up any garden, fitting it for one final hurrah before the slow but, hopefully, elegant descent into autumn.

Weather Watch

August weather has a talent for drama. The pent-up energy of summer heat and humidity can be released in towering, architectural cumulonimbus clouds, which often put on spectacular displays of thunder, lightning, torrential rain and even hail. In August 1879, a ferocious hailstorm destroyed almost every greenhouse in the Thames valley and smashed nearly 4,000 panes in the glasshouses at Kew. In this month, changes in atmospheric pressure result in winds converging from opposite directions that may cause heavy showers. Where the winds meet, air rises, causing clouds to form. This can make it the wettest month of the year on average in some parts of Britain.

WEATHER FACTS

- Average highest temperature in UK: 19.1°C
- Average lowest temperature in UK: 10.8°C
- Highest average regional temperature in UK: London, 23.2°C
- Highest number of hours of sun: south east England, 215.2
- Highest number of days rainfall: northern Scotland, 16.9

WEATHER ALERT

Protect plants from August downpours. The intermittent nature of August rainfall means it's a good idea to capture the water while it's falling so it can be used in the long dry spells between showers. Diverters can be installed in downpipes to direct rain into water butts, which can be connected together for extra storage.

Star Plants

10 OF THE BEST

1. Crocosmia

This plant is not a shrinking violet – it struts. It is a fabulously fiery plant that livens up the late-summer border with its bold, sword-shaped leaves and those glorious flowers that arc over the surrounding plants like cobras being charmed from a basket. And as for that blazing red – it is as intoxicating as a bucket of vodka and as brazen as a squirrel on a bird feeder. Good varieties to try include the bright red 'Lucifer' or bright orange 'Emily McKenzie'. Flowers August to September.

2. Poppy 'Ladybird'

This amazing poppy is as scarlet as a guardsman's tunic with bold black spots. It will never fail to bring joy to even the grumpiest gardener. Most poppies are done by the end of June but not this one, as you can sow it right through the summer – bearing in mind that it takes about 12 weeks to flower. Good in a vase if you dip the cut ends first in boiling water. Flowers June to August.

3. Abutilons

If you're looking for something with an exotic air but that's vigorous and easy to grow, then look no further. Abutilons fit the bill perfectly. These are sun lovers, and vigorous too. None are fully hardy everywhere in this country, but most are good in a large conservatory. However, where these half-hardy shrubs really excel is on sunny fences and walls and in large patio pots, where they bring bountiful summer colour. Some, including varieties of the taller and relatively hardy A. *vitifolium* and A. *x suntense*, are particularly good choices for new gardens, as they grow strongly, flower impressively, then decline as longer-term plantings come into their own, but they're not hardy in colder areas. Flowers May to October.

4. Cannas

The big, bold leaves of jungly plants like cannas are at their best this time of year, making the most of the sunshine. Fast growing with zingy flowers l ate in the season, they bring sub-tropical heat to suburbia. Try 'Phasion' which has fantastic colours in the bronze-purple leaves, striped with pink and tropical-style orange flowers. Flowers June to September.

5. Cactus dahlias

There are many groups of dahlias but cactus and semi-cactus dahlias are among the most varied, the most popular – and often the most dramatic. They come in five different flower size groups: from giant at (25cm+) to miniature (less than 10cm). Giant- and large-flowered forms are the most dramatic, but smaller-flowered varieties integrate more effectively with other plants in summer and autumn borders and work better as cut flowers. Either way, these are dramatic border flowers and splendid cut flowers, whose blooms come in a dazzling array of colours. The huge number of varieties available offers something for all tastes. You can't fail to love the old favourite 'Chat Noir'. Flowers July to November.

6. Canary Island foxgloves

The Canary Island foxglove has long been a favourite at the RHS Chelsea Flower Show – it featured in Andy Sturgeon's 2016 garden. It is such a striking plant that the people looking after the garden were heartily sick of it after five days of people asking what it was called! Flowers July to August.

7. Verbena bonariensis

If *Verbena bonariensis* were a singer, it would be Cliff Richard. Not because it has any particular attraction to a cheesy ballad, but because it just keeps going on and on and on. It will also self-seed if happy, providing you with free plants. Flowers June to September.

8. Rudbeckias

Rudbeckias are really good plants. As perennials for a formal border or prairie planting, as cut flowers, as annuals for summer colour or for punchy containers they're superb. Tough, tolerant, colourful and long lasting – about all they need is sun and not to be waterlogged. And while varieties with traditional daisy-like orange or yellow flowers are certainly impressive, recent developments have brought more colours and new flower forms. There are coppery, rusty or bronze shades, as well as red or even green. Flowers August to October.

9. Red-hot pokers

Red-hot pokers (*Kniphofia*) have striking vertical spikes topped with bottlebrush-like flowers in a spectrum of fiery colours. From red and orange through to yellow and lime green, kniphofia flowers emerge from a clump of long, narrow foliage that's attractive in its own right. They have a long flowering season and look great planted en masse in hot-coloured borders. Flowers June to October.

10. Zinnias

These are vibrant annuals that are easy to grow from seed and thrive in a hot spot. The colourful daisy-like flowers come in vivid colours from red, orange and deep pink to green, with a long flowering season from summer to the first frosts. Use them to fill gaps in the border or create a vibrant container display. Long-stemmed varieties make great cut flowers. Flower July to October.

'This is the best time to plan additions to your August garden. If you can, get out and take inspiration from other gardens. Look around and take actual or mental notes of plants that inspire. Take photos too, both of blank spaces in your August garden and the plants you can imagine filling them. Then, when it's planting time again, this autumn or next spring, you'll be ready and armed with a list of new plants and where to put them for best effect.' – **Carol Klein**

WHY NOT TRY...

Carol's five picks for summer pots

- *Clematis* 'Nubia'
- *Penstemon* 'Sour Grapes'
- *Geranium* 'Rozanne'
- *Dahlia* 'Magenta Star'
- *Heliotropium arborescens*

Jobs to Do this Month

August is a time for holidays, but it's still a busy time in the garden as gardeners try to keep their garden under control. It's a time for dead-heading, more weeding, harvesting and pruning. And for those who fancy trying something new, it's also a good month to save seed and guarantee a good supply of plants for next year. Hopefully, it's also good weather, giving an opportunity to take a break from all the enjoyable labour and admire borders and pots, over a cup of tea.

KEY TASKS

- Cut back rambling roses after flowering.
- Top up bird baths in hot weather when they dry out.
- Start planning your autumn planting to get the best choice of bulbs.
- Trim lavender to keep plants compact.
- Deadhead perennials and annuals to keep them going.
- Stake perennials that are starting to flop under the weight of their flowers.
- Water pots and hanging baskets regularly.

GOT 15 MINUTES?

Top up ponds

Replenish your pond water if levels drop a bit, which is likely due to evaporation. It is preferable to use rainwater. Cooler water adds a welcome supply of oxygen, particularly if you spray it gently onto the surface. If you need to fill it by more than a third, do it in two stages, one day apart, so that any pond life can adjust to the changes in water temperature.

GOT AN HOUR?

Trim hedges

Give a hedge the chance to thicken and harden its new growth ready for winter by pruning it now. Birds will have finished nesting, so won't be disturbed. Use a powered hedge trimmer or pair of hand-held shears. Box, privet and yew can be cut now. Lavender, too, is a good example of a hedge that if trimmed now will keep a good shape for next year. You'll have to cut the flowers, but it is worth it not to have bare, sprawling branches in future.

Save seed

Many border plants are easy to propagate from seed, which allows you to save money and fill your garden with your favourite blooms. Perennials saved and sown this year will make good, strong plants that may flower next summer, while annuals will do so for sure. Leave faded flowers to turn into dry, brown husks, pods or capsules filled with seed. Bear in mind that seedlings of some highly bred cultivars won't be exactly like the parent – colour variations will depend on this year's pollination and the previous plants used in the breeding line.

Seed can be collected and sown straight away or saved until the autumn or next spring. Sown now, they will overwinter as small plants; if sown in spring, perennial plants will need another season to flower. Seed can be stored for at least a year – the harder the seed coat, the longer it can be stored.

1   Snip off some of the old flower stalks on a dry day, when the seed heads have turned brown. Hold them upright to avoid spilling the seed.
2   Put the whole dry stalks in envelopes or paper bags. Include a label with each sample so that they are easily identified if you are harvesting different seed types.
3   Separate the seed from debris with a sieve or blow it off. Some seed heads will release the seed easily, but others will need to be broken up.
4   Label a new paper envelope with the plant name and date before decanting the cleaned seed into it. Seal it then store in a tin or box somewhere cool and dry.

Get rid of bindweed

Trace bindweed down to ground level and use a trowel to lift as much root out as possible. It twines into your plants and you may not even notice it until the large white flowers appear. The bits of root left in the ground will shoot again so ensure that you weed thoroughly. Encourage bindweed up canes to avoid the plant smothering its neighbours, and dig roots out of borders regularly. If you choose to use chemicals, use a paint-on spot treatment with a systemic weedkiller, taking precautions to avoid a windy day; and covering over the plant with a plastic bag to avoid risk to passing insects.

> **TOP TIP**
>
> Raise the cutting height of your mower so grass can grow longer; it will then cope better in hot spells.

Clip lavender after flowering

Trim back your lavender with shears. Clipping it now keeps the growth strong and shape compact. It prevents those central bare patches that so often appear on older lavender specimens. Cut off all the flowers and trim the foliage back to about a third of this current year's growth. You should end with a low hummock-shaped plant. The foliage will resprout in the next few weeks and has time to mature and will therefore be perfectly hardy to withstand the winter.

Deadhead for more flowers

Deadheading is the key to keeping the floral display of high summer lasting as long as possible into autumn. Most plants respond to falling light levels by attempting to produce seed as quickly as possible. By cutting off flower heads as soon as they are past their best you are not only tidying

the display but preventing that process happening. It also provokes the plant to quickly produce more flowers in an attempt to produce the seeds so actually increases the colour and vibrancy of your borders.

The key to effective deadheading is to remember that you are pruning the plant rather than tidying spent petals. Always cut rather than break or pull and make your cut immediately above a bud or leaf – however far down the stem that may be.

Sweet peas are putting energy into setting seed rather than making new flowers. Delay this process by picking every bloom once a week. Use scissors to cut the stems as long as possible and remove seed heads as soon as you see them. This, with regular watering, will extend their season into September.

Clean up your pond

Garden ponds will now be clogged up with plant growth, weed, algae and debris from surrounding plants, particularly trees. Clean it up now to prevent the water deoxygenating as the debris starts to rot in the water. Cut back vigorous plants, thin out the leaves of water lilies and other floating leaved plants, and take out any floating weeds. If you can net out the debris that has sunk to the bottom without disturbing the submerged plants, do that too. Leave it all lying around the pond to let creatures get back into the water, then throw it on the compost heap.

 ## Time to Prune

By now our gardens may be a little frayed around the edges. Spring growth has lost its zing and summer flowers are slowing down. Shrubs and herbaceous perennials may be getting shaggy and the garden looking rather unkempt. But a judicious pruning session can quickly reinstate order and give you time to enjoy your garden while the weather is – hopefully – still on our side. Pruning now can improve next year's flowers and fruits too, especially on trained fruit trees. Hebes and lavender can be shaped and tidied after flowering, and the summer pruning window for plants like wisteria and pyracantha is still open. For some evergreen hedges, such as box, privet and thuja, it's now or never, as the nesting season for birds comes to an end and the cold temperatures that can damage newly cut hedges are still a little way off.

'Remember that pruning is brilliant fun — creating neat shapes, gentle mounds and billows, or just reopening pathways that have become overwhelmed with foliage. There's nothing so satisfying nor so rewarding.' — **Frances Tophill**

HOW TO SUMMER PRUNE WISTERIA

You'll have a wonderful display of flowers on your wisteria year after year if you get the pruning right. The main trick is to prune back the young shoots that grow in the summer after flowering. It's called spur pruning, as you cut back to a short spur to encourage flower buds to form for next year. It is easy to spot where to make the cuts because you can still see the old flower stalks. If your plant is new, tie in the stems that you need to keep to form the framework and cut back the rest of the wispy new stems to five leaves. The foliage is so dense in the summer that it can hide young vigorous shoots growing around the framework, or even up into the eaves of a house. Wait until the winter to tidy it up again. Wisteria can be hard pruned to renovate it if necessary but the flowers may take a year or two to develop reliably again.

- Look for old, bare wisteria flower stalks, hanging behind the dense growth of new shoots that can get quite tangled as they twine around in search of support.
- Platforms are best for working at height. If, however, a ladder is the only option, keep both feet on the rungs and steady yourself before making cuts.
- Trace new twining shoots back to the old flowers and cut back, leaving five leaves on a short stub (spur). Tie in shoots that form the permanent framework.

PLANTS TO PRUNE

- Lavender, after flowering
- Pyracantha – new shoots, to reveal berries
- Rambling roses
- Wisteria
- Espalier apples and pears – new growth
- Bay trees

 # In the Greenhouse

This is a busy point in the year for greenhouse owners – summer crops are reaching their peak and it's also time to think ahead to the autumn and winter. Keep on top of your harvests, while looking out for pests and diseases developing in the warmer temperatures. You can now force bulbs for winter, sow parsley and take cuttings as insurance for tender plants such as pelargoniums.

JOBS TO DO THIS MONTH

- Harvest your indoor crops such as chillies. Use secateurs to snip off cucumbers, aubergines, peppers and chilli peppers as they ripen.
- Pick off yellowing and dead leaves from plants and keep floors and benches swept to reduce homes for pests.
- Remove shading at the end of the month to give plants maximum light.
- Sow seeds of parsley into pots or modular trays for fresh supplies through winter.
- Open the windows on sunny days to ventilate the greenhouse. Temperatures drop at night, so if the greenhouse is not ventilated properly, moisture droplets will gather, encouraging fungal disease to thrive, particularly on plants with soft growth.

DON'T FORGET TO...

- Continue mowing the lawn regularly, with the blades set high if the weather's hot and dry.
- Prune pyracantha now to keep it neatly trained against a wall or hedge.
- Transplant wallflowers sown in June to encourage further root development.
- Carry out lawn scarification, aeration, turf repairs and re-seeding this month.
- Stop weeds from shedding seed by cutting off flowers and seed heads straight away. Your garden won't look better immediately but you will have reduced problems in the future. Avoid putting these in your homemade compost.

Fresh from the Garden

The vegetable garden is beginning to offer up a wide range of harvests from tomatoes, peppers and cucumbers in the greenhouse to courgettes, beans, fennel, potatoes and every kind of root crop, as well as lettuce. The autumn raspberries take over from the summer crop and blackcurrants, blueberries, redcurrants and blackberries are all ready to be picked. It is a delicious – and busy – time.

*'The aroma of vinegar in the kitchen signifies the start of pickling season, when gherkins, jalapeño peppers and nasturtium seed pods are stuffed into jars, then covered in ladlefuls of homemade vinegar, infused with fresh home-grown herbs and spices. Weekly harvests of sun-kissed tomatoes need turning into sauces or salsa before bottling up. With such good harvests this month, I know I'll be grateful in the depths of winter when I can provide splashes of summer from the packed store cupboards and freezer.' – **Rekha Mistry**

WHAT TO SOW

- Radish
- Rocket
- Corn salad
- Spring cabbages
- Oriental vegetables

WHAT TO PLANT

- Salad seedlings

WHAT TO HARVEST

- Beetroot
- Chillies
- Courgettes
- French and runner beans
- Sweetcorn

KEY TASKS

- Harvest and store onions.
- Prune summer raspberries.
- Cut off strawberry runners.
- Cover fruit bushes with netting to prevent birds eating the fruit.
- Prune plum trees.

CROP OF THE MONTH: COURGETTE

Did you know? – The difference between courgettes and marrows can cause confusion. Courgettes are thin-skinned, bred to be eaten small, young and fresh, although the fruits do develop into 'marrow' size with alarming speed if left unpicked. True marrows are thicker-skinned and evenly shaped.

Nutrients – Also known as zucchini, this veg is high in potassium, magnesium, vitamin C, folate and dietary fibre.

Storing – Best eaten as fresh as possible, but keeps in the fridge for several days.

Good cultivars – Climbing 'Black Forest' is good for small spaces. Striped yellow 'Gold Mine' has heavy yields. 'Venus' is compact and heavy-cropping.

Sow – April–June

Harvest – July–October

HERB OF THE MONTH: FRENCH TARRAGON

The wonderful anise flavour of the leaves goes well with fish and chicken, and they also help to digest fatty foods. This herbaceous herb looks graceful in a container, when combined with purple shiso, red orach and marjoram. Plant in well-drained soil in a sunny position. Take 3–4cm soft-tip cuttings in the summer, overwinter and pot up in spring. French tarragon can also be divided in spring. You can grow the annual Russian tarragon from seed, but this isn't nearly as tasty.

GOT 5 MINUTES?

Sow spinach

Sow spinach in 2cm-deep drills and keep watering once germinated to stop plants from bolting. Winter varieties like 'Perpetual', 'Winter Giant' and 'Atlanta' sown now will give you a good supply of leaves right through until the spring. Sow plenty, as they will not replace much growth once picked. In the spring, apply a nitrogen fertiliser top dressing to boost leaf growth.

GOT 15 MINUTES?

Remove bolted veg plants

Leaf and root vegetables are inedible once they have bolted or run to flower, with a tough, new flower stalk replacing the tender edible tissue at the base. Some plants, such as beetroot, bulb fennel, chard and onions, are more prone to do this than others. Cool night temperatures and dry soil are usually the main causes for bolting, although most will do it eventually as part of their lifecycle. Lift out crops if they've bolted and throw them on the compost heap.

GOT HALF AN HOUR?

Plant strawberry runners

Create more strawberry plants by planting runners. Choose a sunny, free-draining site and clear it of weeds. Dig in some rotted manure or garden compost and rake a general fertiliser into the surface. Snip the largest leaves off the runners and trim the roots before planting to encourage vigorous regrowth. Plant with the crown of the plant on the soil surface. Water well and remove any flowers that appear this autumn. Fresh, home-grown strawberries taste sweeter than any you'll buy so plant now and you'll have delicious fruit next summer.

'If you sowed a row of carrots in spring, now is the time to reap the pleasure of pulling them fresh from the ground, with that bouquet rising from the earth and an incomparable taste of sweet, rich delicacy. I often have to wait until May for the soil to be ready to sow mine, making the produce more of a treat when it's finally ready.' – **Monty Don**

OTHER JOBS

Harvest main crop potatoes

When the foliage of your potato plants yellows and dies back, the potatoes are nearly ready for lifting. Ideally, they'll need a two-week period of dry weather in the ground – this hardens them and sets the skin to improve them for storing. If wet weather is forecast, it's best to lift them. Use a fork to loosen the ground, then lift the plant and pick out all the potatoes. Wash or rub the soil off them, then select only the perfect tubers for storing. Put them in paper or hessian sacks and store somewhere cool and dark.

Store gluts

Keep picking to ensure your fruit and veg carries on into the autumn, even if the August harvest appears to be overwhelming. Late varieties of soft fruit provide a welcome supply for summer desserts and preserves. Courgettes and beans must be picked small to keep the plant cropping. Water plants after picking to keep them in growth. If you can't keep up with eating the harvest, share it or start to bottle or freeze it for the winter.

HOW TO PICK TO MAXIMISE YOUR HARVEST

'Every plant's goal is to make seeds and reproduce. So if you come along and remove its leaves, pods or fruits before they mature, it has another go and grows some more. Pick cleverly and you'll get your plants to produce food for you over and over again.' – **Sally Nex**

Pick little and often – Courgettes, beans, peas, herbs
Marrow-sized courgettes and lumpy, stringy beans are no good to anyone. Even worse, plants come to a full stop once fruits get too big. So, every 2–3 days, cut courgettes at 10–15cm and pick over your beans and peas. Trim herbs regularly and they stay leafy, too.

Pick leaf by leaf – Chard, kale, lettuces
Loose-leaf 'cut-and-come-again' crops are such good value – take young, tender leaves from around the growing point and more spring up to take their place. I also pick cos and butterhead lettuces leaf by leaf, instead of cutting whole heads, giving me more salad leaves over a longer period.

Leave stumps to resprout – Spring and summer cabbage, calabrese, broccoli
Once you have cut spring and summer cabbages, leave the stumps in the ground and they will resprout new heads almost as big as the first. And calabrese and broccoli don't stop after that first big central floret – keep them growing and dozens more smaller sprigs appear.

Check sweetcorn

Examine the tops of sweetcorn cobs and when the flower tassels turn brown, test them for ripeness. Peel back the leafy sheath around the cob and press your fingernail into a kernel. If the sap is clear, then the cob is still developing; if it is milky, then it's ready. The lowest cobs will be ready first. Pick by snapping off the cob sharply down the stem. Take the leafy layers off the cob, then either cook it whole to eat as it is or use a sharp knife to cut the kernels off and add to salads, hot dishes and pickles.

 # Rekha Mistry's Recipe of the Month

TABBOULEH

Serves four

You'll need

400ml vegetable stock
250g bulgur wheat
3 tomatoes, diced
1 cucumber, cored and diced
Large bunches of mint, parsley and spring onions
1 tsp cumin
Juice of 1 lemon
100g mixed olives
4 tbsp extra virgin olive oil
Salt and pepper to taste

Method

1 While the stock is heating up in a pan, place the bulgur wheat in a heatproof bowl.
2 Add the hot stock to the wheat and leave for 20–30 minutes until all the stock is soaked up.
3 Place the soaked bulgur wheat in a large mixing bowl, and add the cucumber, herbs and tomatoes. Toss well before adding the cumin, lemon juice, olives and olive oil. Season and serve.

You could vary this salad by adding any of these seasonal ingredients:

- Steamed young beetroot
- Diced radish
- Grated kohlrabi
- Sliced red pepper
- Nasturtium leaves and flowers

 ## Wildlife Notes

The long goodbye to summer begins: swallows and swifts, their young in tow, ready for a journey back to sub-Saharan Africa, despite only being a few weeks old. Only the house martins remain – they may still raise young for another month. Few birds are spotted in the garden now, as most species – after using all their energy feeding chicks – are having a 'summer moult', renewing their feathers. For a couple of weeks, as new feathers replace old, they are vulnerable to predators. So they lie low, skulking in the borders. Many bees are near the end of their lifecycles now – bumblebee queens mate before finding somewhere to hibernate until it's time to found a nest next spring. But some bees are still busy – leafcutters are lining their nests in bee hotels with rose and wisteria leaves.

DID YOU KNOW?

Painted Lady butterflies reach the British Isles from March to June, and then feed and breed. Numbers increase greatly in August, as the second generation appears, before they migrate back to their winter homes.

LOOK OUT FOR...

- Wool carder bees 'combing' hairs from lamb's ear leaves to line the nests of their young.
- Slow worms in your compost bin – avoid disturbing the bin as they may have, or be about to have, young.

- Silvery moths resting during the day on walls and fences, where they blend in with their surroundings.
- Iridescent green rose chafers flying clumsily and noisily into gardens looking for roses to feed on.
- Six-spot burnet moths, blue-blacked with six red splodges on the wings, they breed and feed in meadows, and fly during the day.
- Ladybirds – many species, including the two-spot and seven-spot, will have benefitted from last year's hot summer, and an abundance of aphids in spring. New adults will be on the wing seeking old stems and leaf litter to hibernate in.
- Grasshoppers in long grass. Listen for the males' breeding call, which are made by rubbing their legs against their forewings.
- Bats flitting between trees at dusk. They feed on insects, including mosquitoes and moths.

Spotter's Guide to Caterpillars

Caterpillars (from the Old French *chatepelose* – hairy cat) are the eating and growing stage of the butterfly or moth. A few might be considered minor pests because they nibble prize leaves or vegetables, but most are discreet and secretive, feeding only on their allotted wild plants in quiet corners.

Not all caterpillars are hairy (a defence against being eaten by birds, which choke on the bristles), with many being camouflaged like leaf curls or twigs, or warningly coloured to show that their bodies have stored distasteful chemicals from poisonous leaves.

When fully grown, most descend to the ground and burrow into loose soil to spin a silk cocoon for the chrysalis, in which the earth-bound maggot magically transforms into an aerial adult.

5 CATERPILLARS TO SPOT

- **Garden tiger moth** – 60–70mm. Upper body is black, ruddy brown below, with long, pale hairs all over. The archetypal hairy cat, this 'woolly bear' has a mesmerising undulating gait. Eats a variety of food plants, including nettles, docks and hound's tongue.

- **Cinnabar moth** – 25–30mm. Has yellow and black bars with sparse, long, pale hairs. Shreds its food plants to bare stalks, leading to local population crashes. Feeds on ragworts, including groundsel, and coltsfoot. Sequesters ragwort poisons, so also poisonous.
- **Holly blue butterfly** – 15–18mm. Its squat, plump, silky-green body has tinges of pink. Spring larvae feed on holly buds; summer/autumn caterpillars on ivy flowers, but in urban gardens will alternate on pyracantha and snowberry.
- **Elephant hawk moth** – 85–95mm. A bronze-brown to greenish body, with eye spots behind head and tail thorn. If disturbed, it squeezes front segments into a narrow trunk, inflating eye-spots into a snake-like head. Eats several plants including willowherbs.
- **Drinker moth** – 65–75mm. Looks like a gold and blue blanket, with short tufts of black along back and a frayed white skirt. Likes common grasses such as cock's foot. Named for supposed habit of supping dew drops.

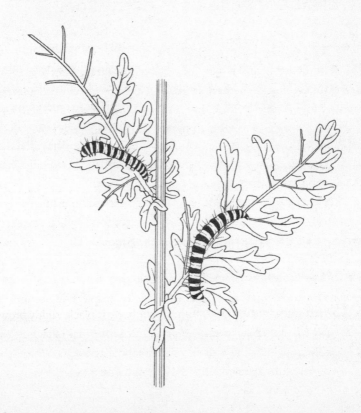

Troubleshooting Guide

RUST DISEASE

There are many different types of rust disease, all caused by different species of basidiomycete fungi. In the UK, around 200 species are known and many of these affect garden plants. In turn, each rust attacks only a small group of plant varieties. So a rust will not spread from your hollyhocks to your roses, for example. A few simple tips cover caring for all types of plants susceptible to rust, including chrysanthemums, fuchsias, pelargoniums, alliums, box, hollyhocks, pears and roses. Avoid overuse of nitrogenous fertiliser, remove and bin dead material in autumn and water carefully. Don't water onto plant leaves, do it early in the morning and ventilate greenhouses well. Rust spreads in humidity. No chemicals are authorised for home garden use against rust on edibles but there are a few for ornamental plants.

MOSQUITOES

The main factor is the number of suitable breeding places where the larvae can develop. If you have any still or stagnant water in containers, you may find the larvae there, so have a look around and tip them out. If you have a water butt with an open top, they may infest this, too – a little cooking oil poured onto the surface will form a thin layer to prevent them from getting the oxygen they need. As long as you don't let the level fall below the butt's spigot, the water you use won't be affected by the oil.

PLANTS FOR NORTH-FACING SHADY SPOTS

Try *Alchemilla mollis*, which has lime-coloured flowers and pleated leaves, *Geranium phaeum* 'Lily Lovell' for its purple flowers and *G. phaeum* 'Samobor' for its purple-marked leaves. *Cyclamen hederifolium*, in its pink or white forms, thrives in these conditions, as does *Digitalis purpurea albiflora*.

Shrubs for dry shade include the evergreen *Daphne laureola*, which has bright greenish-yellow flowers in early spring, and *Euonymus fortunei* 'Emerald Gaiety' for green and white leaves, tinted pink in winter. Plant male and female forms of *Viburnum davidii* for its unusual blue berries.

ANT HILLS ON LAWNS

Put a large biscuit tin or flower pot over the ant hill in the morning. As it heats up, the ants take their eggs up into the container. In the afternoon, slide a piece of cardboard under the container, and remove and dispose of the eggs. They make a tasty treat for birds, especially chickens. Alternatively, remove the container and spread the eggs over the lawn surface for birds to eat. If the surface is uneven due to ant activity, peel back the turf in the raised areas, remove excess soil and relay the turf, ideally in winter when the ants are less active.

The biocontrol nematode *Steinernema feltiae* can be watered into the soil where ants have brought soil up onto the surface. You can also encourage insectivorous birds by hanging bird boxes and feeders.

ROSE BLACK SPOT

Black spot is a fungus that causes dark spotting on rose leaves and stems. Eventually, leaves turn yellow and drop off before autumn. Spores overwinter on the fallen leaves and re-infect the plant in spring when new foliage appears.

Pick up all the leaf litter on the ground and cut off leaves showing signs of black-spot disease. The plant will soon recover and you will have reduced the chance of reinfection.

You can also spread a thick layer of mulch around the base of the affected plant to prevent rain splashing soil-borne spores on to new spring growth. Grow rose varieties that are resistant to black spot – you'll find suggestions in the catalogues of specialist rose growers.

SEPTEMBER

September is a benign month, bestowing a golden grace on all gardens with the gentleness of its light. This can still be summer-warm from the sun but is tinged with the tapestry hues of autumn, and as the month progresses the colours fade and the light gets lower, slanting in through the thinning hedgerows. As a result, the rich colours seem velvety, while there is also a gauzy translucence to the more subtle shades.

But while the flower garden is slipping into autumn, the vegetable garden is at its most productive. September is, above all others, the month of harvest, both of fresh vegetables and those that will store throughout winter. It's a good time to turn your tomatoes into sauce for freezing, so throughout the winter months that lovely taste of summer can be brought alive again. Apples and pears are ripening by the day, and raspberries, strawberries, blackberries and blueberries are still there to be relished.

Weather Watch

Meteorologically speaking, this is the first of the autumn months and we certainly start to notice the shrinking evenings, especially once the equinox has passed, with time for autumn mists to form during the lengthening nights. But very often summer will linger for a while, and it's not unusual for September to have an average temperature similar to June's.

The UK record maximum temperature for September is a scorching 35.6°C measured at Bawtry in South Yorkshire on 2 September 1906. There's many a July and August that have never reached those dizzy heights. However, long-term figures show that it's common for the second half of the month to turn unsettled, with a rise in the frequency of strong winds.

DID YOU KNOW?

Soil retains heat longer than air at the end of summer:

- Air temp: August 17°C • September 14°C
- Soil temp: August 16°C • September 15°C

Based on air and soil (30cm depth) temperatures at Church Lawford, Warks 1981–2010.

WEATHER FACTS

- Average max temperature in the UK: 16.5°C
- Average min temperature in the UK: 8.8°C
- Highest number of days of rainfall, regional average: northern Scotland, 18.2
- Lowest number of hours of sun, regional average: northern Scotland, 80.4
- Highest days of air frost, regional average: East Anglia, 0.6

WEATHER ALERT

Plant and divide herbaceous perennials now. While the soil is holding on to plenty of summer warmth and the shortening days and weaker sun mean that evaporation levels are decreasing, this is a good time for planting out new perennials or dividing clumps. Plants will put on root growth before winter and become better established.

 Star Plants

10 OF THE BEST

1. Caryopteris

In China, the shrub *Caryopteris x clandonensis* 'Dark Knight' is known as 'Bluebeard'. If you look carefully, it is definitely a bit bristly, but to credit it with a full beard might be a bit optimistic. It is a really useful small shrub with lavender-scented foliage and great flowers that (top tip) dry well to be added to winter flower arrangements. There should be room for one in every garden. Flowers August to September.

2. Penstemons

Penstemons have a long flowering season – from June to September, and often almost into winter, those spikes of colourful blooms just keep coming. The flowering stems stand upright, half their length lined with bold, trumpet-shaped flowers, and they happily merge with their neighbours, giving support, creating interesting associations of colour and form, their pointed, glossy foliage always in harmony. 'Schoenholzeri' give a long reliable show of bright pink flowers, or try the hardy, violet 'Sour Grapes'. Flowers June to November.

3. Purple bell vine

If you want a climber with plenty of dangle, then this is the one for you. The flower centres of *Rhodochiton atrosanguineus* hang down like the clappers of a particularly beautiful bell. It is tender, so probably won't last the winter, but is really useful for covering the ankles of other climbers – for example, you know how the bottom few feet of a rose often look a bit scraggy? This is the solution. It's a perennial vine, but usually grown as an annual. Flowers July to October.

4. Fennel

Fennel is one of those plants that bridges the gap between the flower garden and the kitchen garden. It's good as a herb but it's also ornamental and excellent in borders – the stiff stems are particularly useful as an informal support to floppier perennials. The yellow flowers are followed by aromatic seeds which provide food for birds. Flowers July to August.

5. Sedums

Botanists at the RHS have been taking a close look at these hardy perennials for quite a few years and concluded that they're just too different from the other plants named sedums for them all to be called the same thing. The next question was, what should they be called instead? And the answer is *Hylotelephium* – although it's fair to say that some botanists disagree. So, at the moment, the scientific name we'll use is *Hylotelephium*, but it's OK to continue to call them by their common name – sedum. They are all short, rather succulent, sun-loving hardy perennials with noticeably fleshy, blue-grey foliage, often with a misty bloom to it and, in some of the best, deeply flushed in purple. The flat-topped flower clusters tend to open in late summer and autumn, and often change colour from bud to bloom to dried head. They are also among the most attractive flowers to butterflies. What's more, the dead flower heads are often worth leaving on the plants – covered in frost on a chilly morning they look delightful. So, whether you call them sedum or *Hylotelephium*, these late-summer flowers are essential, and easy, autumn perennials. Flowers July to October.

6. Chinese silver grass

Miscanthus sinensis 'Red Cloud' is a more compact variety than most, so good for smaller gardens, pots and front of borders. It has eye-catching red flower spikes and looks good planted with late perennials such as heleniums and rudbeckias. Flowers August to September.

7. Helenium

Heleniums have pretty, daisy-like flowers that bring vibrant colour to the garden in late summer. With blooms in hot yellows, orange and red, they look fantastic planted in drifts through a mixed border. Try companion plants such as taller ornamental grasses, other daisy flowers such as echinaceas, or kniphofias. Heleniums make good cut flowers and are attractive to pollinators. Flowers July to September.

8. Hardy plumbago

Hardy plumbago or ceratostigmas are woody perennials that provide bright blue flowers in autumn. The vivid display lasts until October, by which time the flowers provide a beautiful contrast with plum-coloured autumn leaves. Try the varieties *Ceratostigma willmottianum* or *C. plumbaginoides*. Flowers August to October.

9. Agapanthus

Also known as African lily, agapanthus are perennials native to South Africa. They make an excellent cut flower, with their globe-shaped flowers that come in blues, lilacs and whites. They're also perfect for planting in containers – plant them in spring for late-summer flowers. Deciduous agapanthus are hardier than evergreen types, and can survive British winters if grown in a sheltered spot. Flowers June to October.

10. Nerines

Vibrant nerines provide fantastic colour in autumn, with flowers in shades of pink and red. Nerines are grown from bulbs; most are tender and need to be grown in a greenhouse – the hardiest variety is *Nerine bowdenii*, which can be grown outdoors in warmer regions of the UK. Flowers September to November.

'Although there are lots of different planting styles, for me, a garden should reflect your personality. So, as you get to know your space, don't only think about styles – also consider the mood or emotion you'd like to create. Your planting should not be dictated just by how it looks, but also by how it makes you feel.' – **Adam Frost**

WHY NOT TRY...

Elephant ear

Create a feeling of calm and, hopefully, bring yourself some good luck by adding a dwarf elephant ear plant to your home. The glossy leaves of *Alocasia cucullata* are likened to the ears of elephants, hence its common name. Also known as Buddha's hand, it is believed the leaves wave good luck into wherever it is placed, and for this reason is grown in Buddhist temples in Southeast Asia, where it originates from. In the wild, elephant ear plants can grow up to 7m tall, but a dwarf variety growing in your home isn't likely to get much bigger than 1m. It looks best placed with other luscious-leaved houseplants, such as monstera (Swiss cheese plant).

Position: Good light but out of direct sun. If leaves go pale, try moving it into a shadier position.

Care: Check the soil regularly, it should be kept moist but not soggy. Mist the plant every few days to keep the humidity up if the air in the room is dry.

Details: Keep out of reach of children and pets as it contains calcium oxalate crystals, which are poisonous and irritating. Turn the plant around every few weeks to encourage full growth.

'As summer fades, our gardens can too. We're not the only ones to miss the flowers – it's also bad news for pollinating insects which will be struggling to find food. However, the right container, full of flowers that'll not only look good in autumn but are wildlife friendly, is just the ticket. It's simple to create a sunny display that you can put in the garden, on the patio, or by the front door.' – **Flo Headlam**

FLO'S PLANTS FOR AN AUTUMN POT

- *Libertia ixioides* 'Taupo Sunset' x 1
- *Senecio cineraria* (silver ragwort) x 3
- *Rudbeckia fulgida* 'Goldsturm' x 1
- *Helenium autumnale* 'Salsa' x 1
- *Anemanthele lessoniana* (pheasant's tail grass) x 1
- Container, 46cm

Jobs to Do this Month

KEY TASKS

- Plant winter bedding in gaps in the border or in containers.
- Plant spring-flowering bulbs, apart from tulips.
- Complete more intense lawn treatments like scarification by the middle of the month.
- Move small shrubs and trees if they need repositioning.
- Carry out lawn care before winter.
- Move tender plants indoors.

GOT TEN MINUTES?

Trap earwigs

Protect dahlia flowers from earwigs by trapping them in upside-down straw-filled pots. Release them onto your veg plot.

GOT A MORNING?

Plant bulbs in the lawn

Plant bulbs in the lawn now to enjoy a meadow of spring flowers for years to come. Known as naturalising, this method creates natural-looking colonies of flowers that multiply to give pretty swathes of blooms. These flowers attract pollinating insects and the longer grass gives great habitat for many other wild creatures. You can choose a mixture of flowers or go for a striking mass of one type. Snowdrops, crocus and daffodils do well in early spring, followed by camassia, chionodoxa and dwarf tulips, which make a great display. Plant in distinct areas and mow paths so that you can walk among the flowers. It takes about six weeks from flowers going over until it is safe to cut the grass for the rest of the summer. If you cut the bulb foliage off before the leaves have died back, the bulbs will lose energy and put on a poor display next year.

1 Cut out a piece of turf with a spade, keeping the depth even. Carefully lift the turf and place to one side while you prepare the ground for planting.

2 Loosen the soil with a fork to the depth needed for your bulbs. Most garden soils have plenty of nutrients, but if it is poor, sprinkle on high potash fertiliser.

3 Plant the bulbs nose up so they will be at three times their depth. Place them randomly for a natural look. Level and firm the soil over the bulbs.

4 Gently tamp the turf back down, making sure the edges butt up to each other neatly. Use extra soil to fill any gaps. Water well to encourage the grass to take root.

GOT A DAY?

Sow a new lawn

Do this now while the soil is still warm and the seed will germinate in about a week. Seed is much cheaper than turf so, if you prepare the ground well, you'll save money and the grass will grow and thicken enough for you to mow before the winter sets in. Forking through the soil should be enough but if your soil is heavy and poorly drained, dig first to improve drainage. Leave it to settle a little longer before sowing. Ensure the ground is weed-free and level before sowing to make mowing easier.

AND A FEW OTHER JOBS...

Plant spring-flowering bulbs

Plant spring-flowering bulbs into containers and garden borders. Daffodils, muscari, crocus and hyacinths can be planted now but wait until November before putting the tulips in. Bulbs can be planted into bulb compost and crowded into the pot for a full display. In the garden they'll flower year after year, if you give them space and plant them at three times their depth. Cover the bulbs and water them. Mark their position with a scattering of grit to remind you where they are before the shoots show.

'The great delight of these spring flowers – aside from their reliability – is their lavish display. They bring vivid colour to the garden when border perennials are only just waking up from their winter sleep and, in so doing, they extend the season of interest – and spectacle – in our gardens by several weeks.' – **Alan Titchmarsh**

Take cuttings of tender plants

Cut off strong non-flowering shoots of tender perennials to make cuttings now. Fuchsias, pelargoniums and argyranthemums are all easily propagated from cuttings taken now. Trim the cuttings just below a leaf joint and snip off the lower leaves. Keep just one or two leaves at the top of the cutting. Insert them firmly around the edge of pots filled with gritty seed and cuttings compost. Soak the pots, then put them in a closed bag or covered propagator to root. Air them regularly and expect them to root within a few weeks.

Keep deadheading roses

Keep deadheading your roses, late-flowering perennials in the border and container plants, such as geraniums, to keep them flowering well into the autumn. Snap them off or use clean, sharp secateurs to snip off the old flower heads. Most repeat-flowering plants have more to give before the end of the season. If you've chosen a variety of rose specifically for its winter display of hips, or a flower for its winter seed heads, it's better to leave them alone now, to allow the hips and seed heads to develop.

Harvest chrysanthemums

Cut stems of chrysanthemum flowers now to give you long-lasting flower arrangements in the house. The harvesting season for garden chrysanthemums lasts until November. The multi-flowered early spray types are the first to flower. Wait until the centre flower of the spray is open then cut the stem out leaving a stump of about 20cm. Plunge the stems in a deep bucket of water for an hour or two then recut the base and strip the lower leaves away before arranging in a vase.

Feed your compost

Add deadheaded flowers and trimmings to the compost heap in alternate layers with woody waste. Avoid composting seed heads from prolific self-seeding plants.

Plant new flowering perennials

Prepare ground to plant new perennials now, by removing all weeds. The soil temperature retains the summer warmth for several weeks so there is plenty of time for plants to root out and establish before they become dormant for the winter. Cooler day temperatures and the start of the autumn dews are perfect stress-free conditions for plant establishment. Make sure they are watered if it's dry and put some mulch around them to protect the new roots.

*'An established garden need not be a tired garden, but we gardeners need to be aware that left to its own devices and with no attempt to renew, refresh or reinvigorate, our plots will become samey and a shadow of their former selves. Our intervention, preferably on a consistent basis rather than as a one-off exercise, can make our gardens what they should be – ever developing and dynamic.' – **Carol Klein***

DON'T FORGET TO...

- Move evergreens while the soil is still warm.
- Leave faded sunflower heads for birds to eat.
- Fill gaps in borders with late-flowering perennials.
- Trim floppy stems.
- Trim evergreen hedges before temperatures drop.
- Clear spent crops from the greenhouse.

LAST CHANCE TO...

- Move evergreens that need a new position, while the soil is still warm.

Time to Prune

The nights may be drawing in and mornings filled with dew, but September can be a time for warm, golden days in our gardens, and for enjoying any late-summer flowers at their absolute peak and produce at its most bountiful. It is also time to start thinking about the ways in which we can start putting our gardens slowly to bed for the winter to come. Certain plants will be looking past their best and a little neatening now won't go amiss.

Pruning and cutting back things like hedges and borders can be done easily, now that the birds are well and truly finished with nesting and fledging. The wind has probably knocked down many of our grasses and flowers, and most things will have become rough around the edges. The great news is that taking action now will mean that your plants will look neat throughout winter and won't need urgent attention in spring.

We are coming to the end of the time for pruning most deciduous shrubs – you don't want them putting on soft growth that's easily damaged once the frosts kick in.

PLANTS TO PRUNE

- Birch
- Hedges
- Summer-fruiting raspberries
- Herbaceous perennials
- Deciduous honeysuckle, once it's finished flowering

In the Greenhouse

Now that it's September, shorter days and lower natural light conditions make plant growth challenging, so pull up the blinds and take down shade netting. White shade paint can be left until it's time to clean the greenhouse. Keep the vents open on warmer days to maintain cooler inside temperatures, but close them at night when the temperature plummets. The last of the tomatoes, aubergines, melons and peppers will ripen better with more light. The combination of light and ventilation will help stop autumn rots, but check regularly for fungal disease and remove infected leaves. It's also a good time to take advantage of warm, dry days to tidy and clean the greenhouse. It's much easier to do the job when you can leave plants and equipment outdoors for a few hours.

JOBS TO DO THIS MONTH

- Prepare border soil for winter salad growing as your summer crops stop producing fruits.
- Check stored bulbs and tubers and throw away imperfect ones.
- Prepare for an early flower display by sowing hardy annuals such as nigella into pots now. Harden them off and plant out in spring, when they will flower earlier than anything sown directly into the ground.
- Care for tomatoes – as tomatoes' rate of growth slows, reduce watering and stop feeding. This makes the flavour sweet and more intense. It also stops the fruit from splitting, a common problem at the end of the season. Remove most of the leaves, too, so that the last trusses get maximum light for ripening.
- Tidy and clean, to make your greenhouse ready to overwinter tender plants. If summer crops are past their best, throw them out so perennials that will survive winter with protection can be accommodated. Counter pests by cleaning as you go.

 Fresh from the Garden

Anyone with an allotment or veg plot knows that this is the month of gathering in produce, some to eat fresh, such as courgettes and kale, but much (apples, pears and chillies) to process and store for winter.

The days can still be hot, so ripening is good, but they are becoming drastically shorter with the sun slipping lower and lower in the sky and, although this results in some of the most beautiful light of the year, it means the nights are longer and colder, so growth is slowing right down.

Gather in your harvest. What you can't eat now can be saved for future months. There's nothing better than a freezer full of good things for winter.

WHAT TO SOW

- Japanese onions
- Oriental leaves
- Spinach

WHAT TO PLANT

- Strawberries
- Fruit trees and bushes

WHAT TO HARVEST

- Courgettes
- Kale
- Tomatoes
- Autumn raspberries
- Runner beans
- Chillies and peppers
- Autumn cabbages

KEY TASKS

- Store apples.
- Sow oriental salad leaves.
- Protect brassicas from pigeons.
- Plant spring cabbages.
- Cut out the old canes of blackberries.

CROP OF THE MONTH: AUTUMN RASPBERRIES

Did you know? – Raspberries grow well in Scotland and in the 1950s were transported fresh to London on a steam train known as the 'Raspberry Special'.

Nutrition – Raspberries are a good source of vitamin K and fibre. They also contain a lot of vitamin C, essential for healthy skin and bones.

Storing – Best eaten as soon as possible, but if washed and dried first, can be kept in the fridge for a few days, away from the coldest part. For longer storage, pat dry with a paper towel and freeze.

Good cultivars – 'Polka', for fruit until the first frosts; 'Himbo Top', for strong flavour; 'Autumn Bliss', for large berries.

Plant – November–March

Harvest – September–November

HERB OF THE MONTH: CHIVES

Chives have mild, onion-flavoured leaves – the chopped leaves and flowers make lovely summer garnishes for salads and soups. With purple miniature allium flowers, it's a very ornamental herb plant and loved by bees. Chives will die down in winter, but they're perennial so will come back year after year. They grow well in pots but are best suited to growing in the ground.

GOT 15 MINUTES?

Sow oriental leaves

Sow large areas of hardy oriental salad leaves now for harvesting through winter. Seed mixes usually contain rocket, mustard, pak choi, mizuna and Chinese cabbage. These will germinate within days but be sure to sow plenty as, with few growing days left this season, you need time to grow enough leaves to see you through winter. Either scatter and rake seed into a patch or, if your ground is very weedy, sow them in rows to make weeding easier.

GOT HALF AN HOUR?

Harvest brassicas

Start picking autumn brassicas such as broccoli, cabbages, cauliflower and kale now. Cut cabbages and cauliflowers through the stem at ground level. Lift out the roots and compost them now, or leave the short stump to resprout more leaves to crop in a few weeks' time. Cut off the heads of Calabrese broccoli and work down the plant, taking the smaller sideshoots, too. Do the same with sprouting broccoli, leaving it in the ground to produce a few more sideshoots into autumn.

GOT A MORNING?

Get pickling

Pickling produce is inexpensive and easy to do. It can be a sweet or savoury pickle, depending on the recipe, and you can use no end of different fruit and veg. These jars of delicious preserved goodies aren't just for a ploughman's lunch – they taste great added to soups, stews and curries, and work just as well with fish as they do with meat.

Vinegar with salt, or sugar, are the main ingredients used to pickle your fruit or veg, but you can also add other herbs and spices – such as coriander and mustard seeds, black pepper, ginger, cloves and bay leaf – to bring more flavour. The recipe is for chillies, but you can pickle almost anything, including beans, cabbage, cucumber, damsons, fennel, garlic, onions and pears.

1 Prepare your chillies by cutting them into rings, removing the seeds as you go. Why not use several different varieties of chilli to make a colourful jar? Make sure you are using produce that is firm and as fresh as possible.

2 Place the chillies in a sterilised jar that has a secure airtight lid. First wash the jar in hot, soapy water and place in the oven at 140°C to dry completely. The quantity of chillies will depend on the size of your jar.

3 Heat pickling vinegar in a pan – this will give the chillies a softer texture. If you use cold vinegar they will remain quite crunchy. You may also want to add herbs and spices (see above) to your vinegar while it is heating up.

4 Pour the vinegar over the chillies to the top of the jar and seal the lid. Store in a cool, dark place for at least four weeks before using.

AND A FEW OTHER JOBS...

Pick pumpkins

Harvest your squash and pumpkins now to ripen them somewhere dry and warm. This will harden the skin and lengthen their storing time. In hot, dry conditions this can be done outside but don't let them get damp. If you're hoping to win the biggest pumpkin competition, keep them growing but raise them off the ground to stop them rotting.

Earth up brassicas

Stake and earth up the taller winter brassicas, such as cavolo nero kale and large-headed cabbages. These are likely to blow around in the wind and flop over unless staked, which can damage their delicate roots as well as making it harder to weed beneath them.

Dig up potatoes

Dig out and harvest your main crop potatoes; they've finished growing once the foliage has begun to die back. Use a spade or fork and dig at the base of the ridge, parallel to the row to avoid damaging any tubers as you dig. Once you've lifted a plant, collect the tubers and dig again to check that you have got them all out. Any tubers left deep in the soil will grow next year and increase the danger of carrying pests and diseases into future years.

3 VEG TO START NOW

Garlic 'Picardy Wight'

A strong-flavoured, early maturing, 'softneck' variety. Can be planted at the end of the month.

How to: Plant out in rows, at a depth of up to twice the height of the clove, spacing 15cm apart and 30cm between rows. Needs a sunny spot in free-draining soil. Harvest bulbs in early July.

Lettuce 'Winter Density'

A really hardy lettuce, producing sweet hearts throughout winter.

How to: Sow in module trays, three or four seeds per module, in a cool

greenhouse or cold frame. Prick out, pot up and plant out in the green-house or under cloches. Harvest a few outer leaves as and when, or pick the whole lettuce when the heart is firm and tightly formed.

Broad bean 'Aquadulce Claudia'

The hardiest variety, so start planting at end of this month.

How to: Sow in rootrainers or pots to prevent mice eating the seeds. Or sow direct in a sheltered site 20cm apart, with 45cm between rows. Keep plants well watered, weeded and support as they grow. In spring, nip off the growing tips as these are what the blackfly love. You'll be harvesting as early as May.

Rekha Mistry's Recipe of the Month

CHILLI JAM

You'll need

50g red chillies, cut in half lengthways and de-seeded

1 plum tomato, peeled and de-seeded

2 or 3 dates, stones removed and chopped

1 garlic clove, peeled and crushed
10g ginger, grated
150ml vinegar
250g sugar
100ml water

Method

1 Blitz the chillies, tomato pulp, dates, garlic and ginger, and add a little water to help create a paste.
2 In a heavy-bottomed pan, melt the sugar, vinegar and water over a gentle heat.
3 Add the chilli paste and simmer before bringing to a gentle boil. Stir occasionally to stop the contents sticking to the base of the pan.
4 Cook down until the majority of the liquid has evaporated. Test by moving a spatula across the base – no liquid should be released on the other side.
5 Carefully bottle up the hot contents into sterilised jars. Seal with a lid.

Note: Once a jar has been opened, it should be stored in the fridge, for up to a month.

 Wildlife Notes

Insects are flocking to late-summer flowers. Look out for red admiral, small tortoiseshell and comma butterflies feasting on nectar before hibernating or flying back to continental Europe. Spot clouds of hover-flies and the odd, scruffy, common carder bumblebee, worn and faded from a summer of visiting flowers in sunshine. Now's the time for fattening up and knuckling down. Many queen bumblebees are already hibernating. Frogs, toads and newts will have found shelter in compost heaps, log piles or at the bottom of ponds. If you stopped feeding hedge-hogs in summer, then start feeding them again this month. Look out, too, for any 'autumn orphans' that are too small to hibernate – call your local hedgehog rescue if you see one.

- Ivy bees nesting in clusters in lawns and borders. They typically feed on ivy flowers.
- Toads returning to overwintering sites. They will remain there until it's time to breed again in spring.
- Seven-spot ladybirds congregating in seed heads. Leave these in situ. However, the harlequin ladybird is known for hibernating in large groups in houses – they're easily moved by gently tipping them into a shoe box and transporting them to your shed.
- Common carder bumblebees. These gingery bees are often faded and balding at this time of year.
- Garden (orb) spiders. The females erect huge cobwebs between shrubs and across paths.

Spotter's Guide to Spiders

Most UK spiders have a one-year life cycle, overwintering as eggs hidden inside a silken igloo in a quiet nook, and hatching as tiny spiderlings in spring. Now is the peak month of full-blown glorious adulthood. All spiders hunt invertebrate prey and kill by injecting venom through hollow, hinged fangs. Despite tabloid headlines, only a few have fangs long or strong enough to puncture human skin – but handle spiders carefully and respectfully, or not at all. Most spiders produce silk from the tip of the abdomen, but not all spin the distinctive circular orb webs you see bedewed on cool mornings. Other uses for the silk include trailing a safety abseil line in case of accidents, trussing up prey, covering eggs and even wrapping a gift for a potential mate.

5 SPIDERS TO LOOK OUT FOR

- **Garden or diadem spider** – Body 10–20mm long. It comes in an array of different colour forms, but usually with black-barred legs and a cross-shaped array of white blobs (the 'diadem' of its name) on the abdomen. It spins large, round webs on herbage. Egg-loaded females are huge compared with males.

- **Zebra spider** – Body 5–7mm long, black with white or cream chevrons on the abdomen and speckled legs. Found on walls and fences, it has a jerky agile gait and jumps several times its body length onto prey. It trails a silk safety line in case of rare misses and falls.
- **False widow spider** – Body 6–9mm long, glossy black with pale markings, often like a fleur-de-lis on the abdomen. It hangs upside down in a tangled web in sheds and animal hutches. Found mostly in southern England.
- **Pale crab spider** – Female 8–10mm long, with a white, pale yellow or greenish body that is broad and slightly flattened. Its front two pairs of legs are very long. The male is much smaller, narrower and darker. It sits motionless, camouflaged in a matching flower, and snatches visiting fly prey.
- **Wasp spider** – Body 12–25mm long, with distinctive black and yellow bars. It makes a small web low in rough grass and decorates it with a pale fluffy zigzag of woolly silk down the centre. It preys on grasshoppers and bush crickets, and is found mostly in southern England.

Troubleshooting Guide

CODLING MOTH

If you've bitten into a juicy home-grown apple to find a caterpillar got there first and has left a messy brown tunnel, this is the work of a codling moth. Females lay eggs in summer, then the hatching caterpillars burrow into a developing fruit, often entering through the calyx (opposite the stalk), so you don't see a hole. After munching through, the caterpillars exit and find a crevice in bark, where they spin a cocoon to sit out the winter.

If you can spot the caterpillars hatching before they enter your apples, you can spray trees with a chemical treatment – although this isn't organic, and may harm pollinators. For an organic alternative, target the caterpillars with a nematode spray after they leave the fruit.

Putting corrugated cardboard around tree trunks can catch the caterpillars before they turn into cocoons, too.

Or you could encourage more predatory birds, like blue tits, into your garden to do the job for you.

BROWN LAWN

Like all the world's grasslands, your lawn is resilient and will soon recover. To help it along, scarify it every autumn to remove debris. If it's a small area, use a handheld wire rake to vigorously scratch the whole surface. With medium lawns, use an electric lawn rake or mower attachment. For large areas you may prefer to hire a powered scarifier. Rake one way first, then at 90 degrees. You'll be surprised how much 'thatch' (moss and dead grass) comes out. Afterwards, scatter lawn seed over the whole area, concentrating on any bare patches, then water well. Your lawn will soon revive.

OVERWINTERING TENDER PLANTS

The best place to overwinter tender plants is in a warm but bright location in your home. A windowsill facing east, south or west is ideal. You can cut back some tender plants by half (such as salvias and pelargoniums) before bringing them under cover so they take up less space. While indoors, these plants need just a little ongoing care. Regularly turn the pots to prevent plants growing in one direction as they reach for the light. And water them occasionally, but only when the compost dries out.

NON-FRUITING APPLE TREE

This is a common problem. If an apple tree has a heavy crop one year, it may take a year off to recuperate. The year after this, it will probably flower profusely and crop heavily again. To help it recover, water the tree in dry weather and add a mulch of well-rotted manure or compost around the base. Also apply a dressing of wood ashes – these contain potash that encourages flowering and fruiting. Then at the start of July, thin out the fruitlets ruthlessly, removing every one that is misshapen, damaged, blemished or congested. Leave only the very best, well-placed fruits. This will reduce the burden that year, so enabling the tree to repeat the whole process again the year after. But only as long as you thin out the fruits heavily in early July every year.

OCTOBER

The trouble with October is the months it lies between. The span between September and November – the first being the best of late summer and the latter reaching into the black heart of winter – is an awful lot for one month to accommodate.

There can be glorious days of soft sun, with the garden still flowering profusely. But there can also be sweeping rain and frost, and above all the sense that each good day is torn from the calendar and lost. The soil, though, is still warm, so it's a good time to plant, and the leaves – while they last – range from gold to deep plum.

Weather Watch

October can sometimes give us wistful memories of summer heat when balmy breezes waft in from the south. The dominant leaf and flower colours in the garden gradually move towards the warmer end of the spectrum. And these reds, yellows and oranges are accentuated as the light starts to take on a more golden hue. Although at the start of the month the sun climbs around halfway into the southern sky at noon, by the end of the month it rises little more than 30 degrees above the horizon. One effect is weaker light, but more blue light is also filtered out of the spectrum. This means that the light reaching us directly is much richer in colours typical of a sunset, making that autumn foliage glow.

WEATHER ALERT

Get ready to bring tender plants indoors. Weaker sunshine and longer nights mean we can see the first significant frosts of the season this month, especially in colder parts of the country. Make sure the greenhouse is scrubbed out and tidied up in advance, ready to receive tender plants if negative temperatures appear in the forecast.

WEATHER FACTS

- Highest number of days of rainfall: northern Scotland, 21.3
- Lowest number of days of rainfall: London, 8.1
- Average max temperature: 12.8°C
- Average min temperature: 6.2°C
- Highest number of days of air frost: East Anglia, 4.4

 # Star Plants

10 OF THE BEST

1. Asters

Some plants flower in clumps, some in sprays, some in wisps, some in elegant flushes, others in dots, and some even in gushes and splats. *Aster* 'Calliope' is... a cloud. A veritable stratocumulus of tiny lavender daisies. Asters (or *Symphyotrichum*, as we must learn to call them) sit there patiently all summer, trying very hard to blend into the background – sometimes it feels as if they will never get their act together but, like the US Cavalry, they appear at the last minute to give our gardens the final summer hurrah. Flowers August to October.

2. Chrysanthemums

Time was when every self-respecting gardener grew lots of chrysanthemums. Whether for picking or for competition, they were very popular. Sadly, no more, although maybe they, like the dahlia, are due for a renaissance. They will flower profusely and look good in a vase for over a fortnight. Flowers September to November.

3. Hebes

Flowering hebes add a wide range of colours to autumn borders, from purple to white, through lilacs, blues and pinks. The spikes can be long and impressive and, in the best varieties, open over a long season. Small-leaved variegated types tend to make neater little bushes whose green-and-purple, pink-and-cream or silvery foliage is pretty all year round but develops its finest colouring in winter. For a late-flowering variety try 'Mrs Winder' or 'Caledonia'. Flowers June to October, depending on variety.

4. Wedding cake tree

Cornus controversa 'Variegata' is a tree that lightens dusky corners of the garden with its sparkling variegations and shelters plants beneath its comforting boughs like a wonderful spreading umbrella. It makes a good specimen tree in grass or a medium-sized front garden. Flowers June.

5. Paperbark maple

Acer griseum is a good tree for a medium-sized garden. It's easy to look after, but make sure it's kept well watered, especially when young. Some of the very best for autumn displays are the maples, of which this is an excellent example, with its papery bark and peculiarly pretty three-lobed leaves that colour up nicely – everybody loves a two-for-one offer.

6. Persicaria

From summer to autumn, these low-lying plants provide the prettiest ground cover. You'd be forgiven for not knowing what a persicaria is. Botanists have been shuffling the names for years but, even if the name is unfamiliar, you'll recognise the plants. Those now brought together under *Persicaria* are an unexpectedly varied group. They include bold border perennials, rock garden plants and even one that shows itself off best in summer containers. And while many are grown for their spikes of usually pink or red summer and autumn flowers, some are grown for their leaves. Most, though, are hardy perennials that begin their contribution to the border in midsummer and continue well into autumn. Flowers June to October.

7. Callicarpa

This shrub is also known as the beauty berry and is a particular highlight in the autumn, with striking purple berries that stand out on bare branches. It's perfect for those who want a standout plant for winter. The berries appear from October to November and it also has good autumn leaf colour.

8. Hesperantha

Crimson flag is a wonderful autumn-flowering plant, bearing delicate pink-red flowers as everything else in the garden dies down. The plants resemble small clumps of fragile gladioli with long thin foliage and upright flower-spikes, each topped by several flowers. Flowers September to October.

9. Smoke bush

Cotinus is known as the 'smoke bush', for its soft pink, summer flowers that resemble clouds of soft smoke. The drama though comes from its leaf colour. There's a range of cultivars with leaf colour from green through to purple, and they all have brilliant autumn colour, with a fiery display of yellow, red and deep scarlet. Smoke bushes are mostly quite large shrubs or small trees, but there are small cultivars available.

10. *Rosa rugosa*

We think of roses in the summer, but there are some varieties worth considering for their autumn appeal, too. This is a tough, easy to grow rose that produces large scarlet hips in the autumn. It's a good choice for hedging with fragrant pink flowers in the summer from June to August.

NOW'S THE TIME TO...

Get planting

While many associate autumn with bulb planting, it's also the ideal time for planting shrubs, perennials and trees as well as laying turf. For many of us, autumn is the highlight of the gardening calendar. A time to plant for next year, make corrections and edits to the current year's display and an opportunity to take full advantage of moist and warm soil – conditions that are perfect for establishing plants.

Jobs to Do this Month

KEY TASKS

- Apply mulch to borders.
- Make leafmould.
- Tidy ponds before winter.
- Store dahlia tubers.
- Take cuttings of plants that might not survive winter.
- Plant tulip bulbs (see opposite).

GOT 15 MINUTES?

Plant some bedding

Bedding planted now will last until spring. The choice of plants available has improved, with small shrubs, ferns and perennials added to the mix. Pansies, polyanthus and even ornamental cabbages are the usual choices. Be aware that wallflowers won't flower until March or April. Also try winter-flowering heathers or berried hypericums and skimmias – these could be replanted in permanent positions in spring.

GOT HALF AN HOUR?

Store dahlia tubers

Dahlia tubers can be left in the borders, under thick mulch, only if you are in a mild region of the UK and have free-draining soil. The rest of us have to lift and store them to be sure of their survival. Cut the stems to 15cm before lifting the tubers. Brush as much of the soil off them as possible. Put them in an airy place for a few days to dry completely. Be diligent with this drying stage as it is the best defence against rot. Place the tubers in trays or pots and cover loosely with dry compost, wood chips, shredded paper or sand. Store in a cold greenhouse or garage until the spring.

GOT A MORNING?

Give borders an autumn tidy

Go carefully with the autumn tidy-up, as some plants are better left for the winter. The strong, upright seed heads of echinacea, monarda, agastache and rudbeckia look pretty in the frosts and feed birds. Leave the stems of tender shrubs like salvia to protect emerging spring growth. Clear away the old leaves of hosta, acanthus and day lilies, and rake off anything else that yellows and turns to mush in the frosts. Tie back any climbing plants that could suffer in the wind, and trim back the softest growth on deciduous shrubs like roses. Weed through as you go, and finish off by mulching.

AND A FEW OTHER JOBS...

Plant tulip bulbs

Tulips also grow well in borders, planting the bulbs 15cm apart and 10cm deep. Six to eight weeks after flowering, dig them up and dry them off, keeping the largest bulbs to replant next autumn. If you want them to come up every year in the same spot, then plant them 20cm deep – the greater depth helps them to survive. Tulip bulbs don't begin to root quite as early as daffs, so it's best to plant them a little later, during October or November.

1 Choose large bulbs, as these have the greatest flowering potential. Make sure they are fat and firm. Avoid any bulbs that show signs of a greeny-grey fungus.

2 Select a large pot, to create a high-impact display. It will also dry out less rapidly than a small pot. Ensure it has drainage holes and use good-quality multipurpose compost.

3 Fill the pot to within 10cm of the rim and space the bulbs 10cm apart. This gives them room to grow, yet ensures they are dense enough to create good floral impact.

4 Plant up the top with bedding to provide interest through winter, before the tulips appear – choose wallflowers, dwarf double daisies or winter pansies.

*'Of all the things I look forward to in the garden, the tulip season ranks high on my list. It's the first burst of proper colour in the garden after the white of snowdrops and the yellow of daffodils. The sheer range of colours and combinations possible means I can create a different show in pots every year.' – **Alan Titchmarsh***

ALAN'S 5 FAVOURITE TULIPS

- **'Abu Hassan'** – Rich-crimson petals, edged with yellow
- **'Estella Rijnveld'** – Blowsy red-and-white 'parrot' type
- **'Prinses Irene'** – A wonderful blend of orange and plum
- **'Queen of Night'** – The best deep purple – almost black
- **'Spring Green'** – Cream goblets, flushed with green

Look after your lawn

Continual mowing takes its toll on a lawn. Take time to weed and scarify by scratching up the moss with a rake or machine. Put the debris on the compost heap, providing you've not used any herbicides. Improve aeration using a fork. Repair worn patches with new turf or seed and lift sunken edges by cutting in and packing new soil under the edges. Keep mower blades high until next year.

Mulch your borders

A surface layer of compost, bark or gravel will keep moisture in the soil, reduce weed germination and, in the case of composts, slowly decompose to improve the structure and add nutrients. After the autumn clear-up, weed borders then spread a 5cm layer of mulch after rain. Bark chips are good under trees and shrubs, but don't work them into the soil as they take years to decompose. For soft-stemmed perennial flowers and veg gardens, compost is best.

DON'T FORGET TO...

- Reduce watering of containers and pots, keep them moist rather than wet.
- Cut back tall shrubs, such as buddleia, to avoid damage by wind-rock in winter storms.
- Bring frost-tender plants inside to protect them over winter.
- Remove leaves from under roses to protect them from fungal problems like rust and black spot.
- Mulch borders.

 ## Time to Prune

This month, focus on keeping your borders looking great for autumn and winter by getting them into shape. Collapsed, soggy or dying-back foliage should be removed to prevent the build-up of slugs, snails and diseases. Plants such as campanula, helenium and sidalcea will be looking their worst, but pruning back this month will result in a healthy plant that will produce colourful flowers next year. Keep in mind, though, that lavender and rosemary will not do well if they are pruned now.

Resist the urge to prune deciduous trees and hardy shrubs, leaving them until midwinter or very early spring, when the plants are still dormant. This will discourage tender new growth from appearing during cold spells, weakening the plant. Hardy yew is one exception, though. Now is the perfect time to give yew hedges or topiary a final prune before winter, so they keep their neat shape.

'With summer a fading memory, now is the time to expect the first frosts, cobwebs and autumn colour. And with a regular pruning regime, you not only keep your garden tidy, but you also create dramatic effects with silhouettes and seed heads – and remember that brown is a colour to be enjoyed, too.' – **Mark Lane**

PLANTS TO PRUNE

- Hardy geraniums
- Virginia creeper
- Buddleia – trim by a third to prevent wind-rock (prune again in spring)
- Blackberries
- Blackcurrants
- Summer-flowering jasmine

In the Greenhouse

The trickiest decision at this time of year is knowing when to call time on summer crops in the greenhouse. Tomatoes, cucumbers, peppers and aubergines are starting to slow down, but they'll keep trying to fruit until the first frosts if you let them. These crops are unlikely to ripen properly, though, so it's generally time to call a halt around mid-October when plants lose that lush, jungly look and start looking gaunt and tired.

Things to move under cover this month include summer patio containers, packed with annual herbs and salads. Under glass, it's three to five degrees warmer than outside, enough to give baby-leaf salad, coriander, flat-leaved parsley and basil a second wind.

This lift in temperature is often enough to keep plants such as frost-tender herbs, citrus and scented-leaf pelargoniums happy without any expensive extra heating. As long as you keep the compost dry, these plants can withstand temperatures as low as -5°C under cover. During colder snaps, cover your crops with temporary tents of newspaper or hessian, removing them once the weather improves.

Warmer winters mean you can't rely on cold weather to kill off pests and diseases. Greenhouses are their favourite places to hunker down, so it's up to you to clear them out before they get too comfortable. If you skip this step in the winter, you'll find aphids, whitefly, slugs and fungal diseases like botrytis will emerge earlier than ever next spring to reinfest your new crops.

JOBS TO DO THIS MONTH

- Pull up the remains of this summer's plants and sweep up any dead leaves left behind, as they can harbour fungal disease. Take it all to the compost heap. Then have a good clear-out, removing empty pots, plant supports, spent grow bags and old compost – all popular hibernating spots.
- Give the whole greenhouse a good wash, cleaning the glass until it sparkles – you'll get more light through clean glass too. Run a plant label along the joints between glass and frame to winkle out moss and other debris where bugs may be lurking, then scrub down greenhouse staging with disinfectant. Your greenhouse should now be sparkling like new and safe to plant again!
- Check your plants for mould and signs of pest damage too. Remove rotting material and banish pests by washing them off with soapy water or a soap-based insecticide.
- Make sure your greenhouse is warm enough for the plants you're keeping there over winter. If you have heaters, check they work.
- Pot up amaryllis (*Hippeastrum*) bulbs to enjoy the impressive flowers in about eight weeks, in good time for Christmas. Keep in a heated greenhouse in maximum light. A bright indoor windowsill works well, too.

Fresh from the Garden

The vegetable garden can still be very productive at the start of the month, with the shortening light limiting growth as much as the temperature. In the first half of the month you can still harvest veg such as lettuces, rocket, mizuna, runner beans, squashes, fennel, spinach, chard and beetroot as well as root veg. Early October is not too late to sow some grazing rye as green manure and, if you have a greenhouse or tunnel, lettuces, mizuna, mibuna, corn salad or rocket can be planted into beds that were occupied by tomatoes, to provide fresh salad leaves until next spring.

WHAT TO SOW

- Broad beans
- Peas

WHAT TO PLANT

- Garlic
- Onion sets
- Spring cabbages
- Blueberries

WHAT TO HARVEST

- Autumn cabbage
- Beetroot
- Carrots
- Potatoes
- Pumpkins
- Swedes
- Turnips

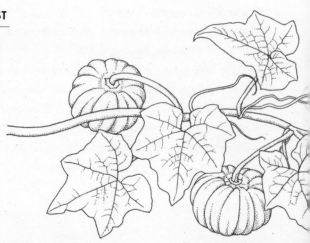

KEY TASKS

- Plant garlic.
- Remove old tomato plants indoors and out.
- Sow green manure.
- Harvest and dry chillies.
- Clear plot of all weeds.

CROP OF THE MONTH: PUMPKIN

Did you know? – While garden-swamping giant pumpkins grab the headlines, there are better-behaved and compact varieties, including neat little miniatures that can be grown in containers.

Nutrients – Especially high in vitamin A and a good source of vitamin E, folates and potassium.

Storing – Whole fruits can be kept for up to two months in a cool, frost-free place. First, toughen the skin by 'curing' – placing in a dry, sunny place for 10 days or so. Otherwise, peel, cook and freeze.

Good cultivars – Small pumpkins, but vine is normal size for small gardens; 'Jack Be Little', trailing / climbing, small, orange fruits; 'Amazonka', compact, semi-bush habit; 'Pot of Gold', early ripening, compact habit.

Sow – April–May

Harvest – September–October

HERB OF THE MONTH: LEMON THYME

Lemon thyme is a very useful herb, as the flavour of the leaves goes well with both vegetable and meat dishes, and is excellent for marinades and sauces. Thyme is nature's natural antiseptic. All varieties grow well in containers, especially when you add extra grit or perlite to the compost, to enhance the drainage. It looks great planted with oregano, sage, savory and prostrate rosemary. Take cuttings in early summer before flowering, or grow from seed, though variegated forms must be grown from cuttings.

GOT 15 MINUTES?

Plant garlic

Choose soft-neck varieties of garlic like 'Solent Wight' or 'Germidour' as they dry and store better than the hard-neck types. Garlic does well in manured ground, but add lime if you have acid soil. Separate the cloves, then plant in rows 30cm apart, allowing 15cm for each one. Plant pointed end up, with 2–3cm of soil above the tip. They will shoot in the next week or two, but most of the growth happens in spring. Water in a dry spring to help the bulbs to swell, but otherwise they need little care until the foliage begins to die back in June. Lift and dry the bulbs in the sun or a cold greenhouse.

*'Garlic has a home in everyone's kitchen, no matter what sort of food you like to cook. If you haven't grown it before, now's the time to give it a try. There are many different varieties out there – all with subtle flavour differences – giving you lots more choice than you'll find in the supermarket.' – **Adam Frost***

GOT HALF AN HOUR?

Plant spring brassicas

Cabbages, spring greens, kale and sprouting broccoli can all be planted now for a crop next spring. Plant into firmed ground right up to the first set of leaves so that when they grow away they have a sturdy stem for support. They can also be earthed up later if necessary; brassicas will produce more roots from covered stems. Be sure to give them plenty of space because they put on a massive amount of growth as the days lengthen in the spring. You'll be picking these tender greens by March.

GOT A MORNING?

Go on a pest hunt

During the autumn, pests find safe places to overwinter ready for next year's breeding cycles. You'll find snails hiding along raised bed walls and behind climbers. Disturb the ground to give birds a chance to find pest larvae and feed on slug eggs. Put greasebands on the base of your fruit trees to stop winter moth caterpillars from climbing up the trunk to nest in the tree. Install some bug boxes or position bunches of twigs to encourage aphid predators like ladybirds and lacewings. Keep the garden birds fit by feeding them through the winter. They'll stay near, nest nearby and keep the bugs at bay next spring.

OTHER JOBS

Clear old crops from your plot

Lift the last of the summer crops out of the ground so that you can get the beds ready for next spring's sowings. Pull out the plants and any remaining weeds. Clean the netting and support canes before storing for the winter. It's a good strategy to cover the soil with compost then cover with dark sheets for weed control. Lift them off in the spring to reveal clean soil and a probable population of slugs that will have gathered on the sheet cover and can be removed.

Store fruit

Go through the fruit you've harvested and select the perfect, unblemished ones for storage. Pears and apples will keep for several weeks if they are stored somewhere cool, dark, airy and dry. They are best separated, so either space individually on a rack or wrap each one loosely in newspaper. Check your fruit store regularly and remove any showing signs of rot.

 # Rekha Mistry's Recipe of the Month

A TWIST ON RASPBERRY CHEESECAKE

You'll need

220g cottage cheese
200g full fat soft cheese
125g caster sugar
140ml soured cream
2 tbsp cornflour
⅛ tsp cardamom powder
⅛ tsp grated nutmeg
¼ tsp vanilla extract
3 eggs
150g fresh raspberries
few strands saffron
For biscuit base
250g digestive biscuits
50g butter, melted

Method

1　Heat the oven to 160°C (fan). Crush the biscuits, mix in the butter and press the mixture into a greased, 20cm loose-bottomed, round cake tin.

2　Whisk together cheeses and sugar, followed by soured cream, cornflour, spices and vanilla.

3 Add the eggs one at a time, whisking the mixture in-between.

4 Fold in the raspberries, and crush and scatter the saffron.

5 Pour onto the base and bake for an hour or until it is set with a slight wobble.

6 Turn off the oven and leave the cake inside for an hour before removing. Let it cool, then chill in its tin in the fridge overnight. Bring to room temperature and serve.

 ## Wildlife Notes

Everything feels very autumnal, suddenly. There are goldfinches on seed heads, light frosts and migration. Frogs, toads, newts, hedgehogs and slow worms are all hunkering down to hibernate over winter, while redwings and fieldfares arrive to feast on windfall fruit. Now's the time to change the food in your bird feeders – give them a good clean before filling with calorie-rich suet treats, seeds and nuts that will help birds find the energy for the hard times ahead. Any late flowers will be providing the last of the year's nectar for straggling pollinators. Many of them, too, will be looking for shelter in the coming weeks, so spare them a thought when you come to tidy the garden. Hedgehogs will need feeding right up until hibernation, so continue providing cat and dog food for them until it's no longer taken.

LOOK OUT FOR...

- Shifty-looking blackbirds and robins. These will have arrived from Scandinavia and may not be as used to humans as those resident in the UK.

- Hedgehogs out during the day – call your local hedgehog rescue for advice.

- The last of the common carder bumblebees. Workers will be faded and worn by now, some almost bald.

- Red admiral butterflies flying back to continental Europe or Africa for winter, although some are increasingly staying here, taking advantage of our milder winters.

- Garden spiders that spin webs across shrubbery in autumn to mate, before both the male and female die, leaving just a sack of fertilised spider eggs to make it through winter and hatch in spring.
- Queen wasps feeding on the last of the ivy flowers, before hibernating beneath bark.

WILDLIFE PROJECT: HOW TO MAKE A PUMPKIN BIRD FEEDER

If you're making pumpkin soup or pie this month, don't waste the tough outer skin – put it to good use and turn it into a bird feeder. This simple design will attract a variety of garden birds, which can nibble at the exposed pumpkin flesh as they dine on seeds.

You'll need

Pumpkin

Sharp knife

Wire

Bird seed

Tablespoon

Method

1. Cut the pumpkin in half and use a tablespoon to scoop out as much flesh as possible, creating a deep pit in one half. Use the flesh in soups or stews.
2. Pierce a hole in the pumpkin half, at the opposite end from the pit. Gently loop the wire through and use this to hang your pumpkins from a tree branch.
3. Holding the pumpkin half in one hand, gently tip a quantity of mixed bird seed into the pit. Gently release the pumpkin so it hangs freely but doesn't swing.

 Spotter's Guide to Snails

Although despised by many gardeners for damaging plants, most snails are secretive, small and harmless in the garden – recycling dead leaves, feeding in soil or on mould, fungus and lichen. On close inspection, you'll discover most shells spiral anti-clockwise on the right-hand side.

Snails mostly feed at night or when it's wet, and hide during the day. In hot, dry weather they go into a dormant state by shrinking inside their shells, which are sealed with a mucus plug. They then turn their metabolism down until autumn.

5 SNAILS TO SPOT

- **Garden** – Large (12–20mm in diameter), globular, prettily stippled with brown and yellow speckles and broken lines. Spread by human activity over numerous parts of the globe.
- **Hedge** – Large (12–18mm in diameter), globular and striped with a variety of broad or narrow bands, but always with pale rim around the opening. Found under logs and in leaf litter.
- **Round** – Small (4–5mm in diameter), and almost flat. The dent (umbilicus) on the underside of the shell is large and deep. Whorls are ribbed when viewed under a hand lens. Found under stones and logs and mostly a soil-feeder.
- **Common chrysalis** – One of several species with a small (3–4mm long), but squat, broad shell, and with the shell opening formed into a distinct broad lip. Often found under ivy, especially on stone walls.
- **Common door** – A tight spire (7–12mm long), with a dark colour – brown to black. Often found on tree trunks. Named after the tiny hinged, round, trapdoor portion of the shell, which flaps shut when the soft body is pulled inside.

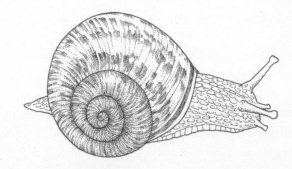

Troubleshooting Guide

LEAF MINERS

If leaf miners attack your plants, you'll find pale wiggly lines across the leaves, with brown marks where damaged areas of leaf have died off. 'Leaf miner' is a broad term for a gang of insects (flies, beetles or moths) that make tunnels by feeding in between the upper and lower surfaces of leaves. Larvae emerge from eggs laid on foliage, burrow in and pupate, before hatching into an adult. Hold up a mined leaf to the light and you'll often see the larva or pupa inside.

In gardens you might find astrantia, chrysanthemum, horse chestnut, beetroot or cabbage leaf miners, among others. Evergreens, such as holly and hellebores, keep their leaf miners with them all winter, but deciduous leaves drop to the ground, leaf miners inside, remaining there until spring. You can sweep up and destroy (not compost) infected leaves, and use specific biological controls to safeguard your plants. These methods are far from guaranteed, but putting fleece around chard and cabbages is a good safeguard if you've had this problem before.

WIND-ROCK

Cut back plants that are prone to wind-rock, such as shrub roses, lavatera and buddleia, by a third to one-half of their height this month and bring vulnerable ones like olive trees indoors if you can. Pick fruit before it's blown down and put mulch on the ground to stop water loss from the roots and soil.

PLANTING BULBS AND PLANTS

The rule of thumb for planting bulbs is 2–3 times the depth of the bulb. This is fine for a bulb you've forgotten the name of, or bought in a sale without a label, but it's worth checking online or in a book if you're unsure. A bulb such as a crown imperial needs at least 30cm depth or it will not flower reliably.

With woody plants, the root flare (the point where the first roots emerge from the stem) should be placed at soil level. This is especially essential with trees as planting them too deeply is a common cause of death. If the root flare isn't visible, scrape the compost back until you can see it.

STAKING TREES

If a tree is stable after planting, don't stake it. If there is a chance of the wind loosening it in the planting hole, insert a low stake at a 45-deree angle, ensuring the prevailing wind blows the stem away from the stake as opposed to knocking into it. A well-placed stake should allow the stem to flex in the wind but the rootball to stay stable. In most cases, stakes can be removed after 12–18 months. If the tree is still unstable after this time it could be establishment failure, typical with planting root-bound stock.

NOVEMBER

November is the month when leaves both strike their best colours and – often just a few days later – fall with real intent. The leaf-mould that they so effortlessly transform into is so useful, both as part of a potting compost mix and as a mulch.

It is time to tidy up and hunker down. Dahlias and cannas need to be lifted and stored, hedges trimmed and tender plants in pots brought in for winter protection. Autumn is declining into winter. The weather shifts into a lower gear and the days visibly dwindle so that by the end of November we can have frost, snow, rain and wind and grey, thin light that fades completely just after four in the afternoon.

Weather Watch

Across all four of the home nations, November tends to be a cold and wet month. With large upland areas in the frontline of moisture-laden air from the Atlantic, the west and north receive more rain than the south and east. North-west England has a reputation for wet weather. In a four-day period, 16–19 November 2009, 495mm of rain fell from an atmospheric river of moisture-laden, south-westerly winds that fed continuously into the fells above the Borrowdale valley – an English record!

WEATHER FACTS

- Average November rainfall in England, 88.2mm; Northern Ireland, 112.5mm; Wales, 162mm; Scotland, 166.3mm; UK, 121.2mm
- Average max temperature: 9.1°C

- Average min temperature: 3.3°C
- Most hours of sun: London, 72
- Least hours of sun: northern Scotland, 23.5
- Highest number of days of air frost: East Anglia, 8.6

WEATHER ALERT

Heavy rain is increasingly likely this month. The organic matter in your soil is crucial as it catches rainwater and allows it to soak in gently rather than wash away your soil or run off to cause a flood. The quickest way to add organic matter to the soil is with a 5–10cm mulch of garden compost. Worms will slowly mix this into the soil to improve the growing conditions for your plants next season.

 Star Plants

10 OF THE BEST

1. Winged spindle

Euonymus alatus is dressed up in its autumnal finest by November – the colour is of a deeply bashful blush, apples fit for princesses and freshly kissed lips. Its leaves are dark green but turn bright red in autumn – for the best colour plant this shrub in full sun. Its fruits are a remarkable pink with orange seeds. Flowers May to June.

2. *Skimmia japonica* 'Rubella'

This is a great evergreen backbone for winter containers. At this time of year the main attraction is the striking red buds. These provide impact in a pot all through the winter, perfect for bringing a hint of festivity to a doorstep, until in the spring they open to white flowers. Flowers April to May.

3. *Pennisetum*

Pennisetums are attractive perennial grasses from warmer parts of the world with their peak season of interest in late summer and autumn.

What sets them apart from other grasses is the mass of bristly bottle-brush spikes, fountains of them, upright or arching, held above slender, graceful foliage emerging from tough, tight, slowly spreading crowns. The flowers may be creamy or various tan and biscuit shades, some with pink or red overtones, with many retaining their looks right through the autumn into winter and even into spring. As features in autumn containers, in drifts through informal prairie-style borders, as intriguing softeners in bold, late-season displays, in gravel... *Pennisetums* have many uses. They're subtle but exude real presence. A good late-flowering variety is the 'Rubrum' which has red flower heads and is best in a sheltered spot. Flowers August to November depending on variety.

4. Cardoon

Cardoons with their massive leaves and towering flower stalks, are a real statement plant for a large border or standing on their own in a front garden. The thistle-like flowers rise to 3m tall and seem to provide enough nectar for every bee within shouting distance. By this time of year the colour has gone, yet this plant still stands like an Assyrian temple worn down by time and weather, battered but unbowed. Flowers June to September.

5. *Cornus* / dogwood

Cornus (or dogwood if you wish to be informal) is always reliable in winter – bright scarlet stems shining in the sunshine. Prune the stems in late winter or early spring, as the younger shoots provide the best winter colour. Flowers May to June.

6. Euonymus

Euonymus is one of those invaluable shrubs that we find in almost every garden – but that, on the whole, no one really gets very excited about. But perhaps we should. These are tough, varied and, in a surprising variety of ways, very attractive shrubs. The evergreens, mainly *Euonymus fortunei* and *E. japonicus*, are simply indispensable. These are the ones we see so often. Always well behaved and available in groundcover, bushy or quietly climbing forms, they are small garden essentials for their variegated foliage and pink winter tints. Flowers May to June.

7. Checkerberry

Nature abhors an empty container. The annuals are over and the bulbs are still underground, girding their loins for next year, so you need something a bit cheery to bridge the gap. *Gaultheria procumbens* is a good evergreen to keep pots looking perky until spring. Pink or white flowers in May and June are followed by plump scarlet berries.

8. *Liquidambar styraciflua*

There is a particularly handsome liquidambar in the RHS Garden Wisley. So handsome is this tree that it's worth visiting at this time of year just to see it. Why? Because of the depth of colour in the foliage – in those dying leaves you'll see shades of rubies, mandarins, fine wine, hessian, log fires, pomegranates, sunsets over tropical beaches, roasted carrot and the blushes of maidens. There are many different varieties available, including columnar forms that would be better for smaller gardens. For the best autumn colour, plant in full sun.

9. *Pittosporum*

If you're strolling down the street and you notice an interesting foliage shrub you don't recognise in somebody's front garden, the chances are it's a *Pittosporum* – they've sort of crept up on us. Once thought too tender for many gardens and grown mainly in cities and suburbs where the climate is a little milder, our winters have become less murderous and hardier varieties have been introduced, so they've steadily become more widely available and more popular. And deservedly so. These are easy-to-grow evergreen shrubs for sunny sites. The well-branched *Pittosporum tenuifolium* is the most widely grown species and it's the foliage that is the foremost feature. Its slender, black twigs carry small, wavy leaves, which can be purplish, bronzed, creamy, silvered or variegated, and among them small, dark, honey-scented flowers open in spring and are especially fragrant in the evening. *Pittosporums* are versatile. Equally good in mixed borders as they are as specimens, informal hedges and container plants in large pots. Flowers May to June.

10. Pyracantha

It's all about berries and autumn colour this month, and pyracantha is a wonderful shrub for a berry display that will brighten your boundaries. Train this shrub against the wall to enjoy the sight of its yellow, red or orange berries through the autumn. The birds will thank you for it, too. Flowers May.

'Everyone, even non-gardeners, knows about autumn colour – it is everywhere. In the heart of the city, the colours of chestnuts, planes, sycamores and cherries wash the cityscape with the tones of their glowing foliage. Those who look to the ground are just as aware of it as those with their sights set higher, as leaves tumble down to create carpets of colour on the pavement. Autumn colour is all a result of our deciduous trees settling down into their winter routine and abandoning their leaves – it's a sign of imminent expiry, which makes it all the more poignant.' – **Carol Klein**

> ### CAROL'S 5 PICKS FOR AUTUMN COLOUR
>
> - *Cotinus* 'Grace'
> - *Hydrangea quercifolia*
> - *Sorbus* 'Joseph Rock'
> - *Viburnus opulus*
> - Katsura tree

Jobs to Do this Month

The focus this month is getting ready for winter – whether that's protecting plants, looking after wildlife or harvesting root crops. It's not too late to plant tulips and, as long as the ground isn't frozen, it's the perfect time to start planting bare-root hedges, roses, trees and shrubs. So, while the days are getting shorter and many plants have retreated underground, there is still plenty to do, always looking ahead to the next season.

KEY TASKS

- Check tree ties aren't too tight before winter.
- Clear greenhouse guttering.
- Clean out bird boxes for next spring.
- Lift dahlia tubers and store.
- Clear faded sweet peas and other annual climbers.
- Finish planting bulbs and bulbous perennials like *Eremurus* (foxtail lilies).
- Wrap tender plants that are staying outdoors.

GOT 15 MINUTES?

Cloche hardy annuals

Protect seedlings sown this autumn, such as nigella or marigolds, before frost sets in. A cloche made of plastic or light fleece will prevent rain and cold winds from causing damage throughout winter. This will not only prevent young plants from getting soaked, but also the small amount of heat retained by the material should keep plants warmer. You will need to water the plants occasionally by hand once they're covered up. Any sown into pots should be brought inside.

GOT HALF AN HOUR?

Wrap tender plants in fleece

If you grow exotics in pots, but don't have a frost-free place to over-winter them, protect them with fleece now. Banana plants, cannas and tree ferns all need winter protection. The crown of the plant is the most important bit to protect, so peg or tie the fleece so that it covers the canopy. The roots are also vulnerable in freezing conditions, so wrap up the pot in plastic bubble wrap.

GOT A MORNING?

Plant winter flowers

Fill any gaps in your borders with winter-flowering plants such as hellebores, heathers, sweet-scented evergreen *Sarcococca* and bulbs like snowdrops that can be bought in pots for instant colour.

1 Use a trowel to make a hole a little deeper and wider than the root-ball. Add some garden compost, then soak the base of the hole.
2 Water the plant before taking it out of the pot and place it in the centre of the hole.
3 Push the soil back around the rootball and firm well to avoid it lifting out in a frost.

A FEW OTHER JOBS...

Make leafmould

Leafmould makes the best soil conditioner but it takes two years to form. The rotting fungi and bacteria need moisture and air so open heaps are best, but bagged leaves work well in a small space. Rake or blow leaves from lawns and paths, but leave them on the borders to rot down natu-rally. Fill a robust black polythene sack with leaves, press them down and water if they are dry before tying the top securely. Puncture the bag with a fork to allow some air exchange as the leaves rot.

Plant flowering shrubs

Prepare the ground by removing weeds and fork plenty of organic matter into the base of the planting hole. Make sure the hole is twice as wide as the pot and the same depth. Firm the shrub in well and keep the woody stem at the same level as it was in the pot to avoid it rotting. Soak with water and mulch with a thick layer of compost but keep the compost away from the stem. Adding shrubs to your border will give it more structure for next year.

Time to Prune

Winter is a peak time for pruning your fruit trees. With deciduous trees and shrubs now devoid of their foliage, berries, coloured stems, evergreens and the seed heads of grasses and other plants are seasonal highlights. It's important to leave all of these unpruned so that they can provide the garden with every ounce of display until the end of winter. However, it's now easy to see where to prune stems and branches of fruiting trees and shrubs that have lost their leaves. Gardeners can pinpoint broken stems, crossing branches and even individual buds, which makes it an easy job to keep shrubs in shape and fruit at its most productive. Most fruit is pruned twice – in winter and summer – winter being the most important time to make cuts, once fruit bushes and trees are dormant.

PLANTS TO PRUNE

- Apple and pear trees
- Grapevines
- Blackberries
- Gooseberries
- Kiwis

🏠 In the Greenhouse

This month, the days are cooler and it's a good idea to use a max-min thermometer to monitor temperatures so you know when to turn on the heaters. Heaters are a good way of protecting tender plants. You'll just need heat at night to begin with. Electric fan or tubular bar heaters are the easiest if you have power. If not, the options are paraffin or gas.

The combination of low light, cold temperatures and high humidity after watering makes the perfect conditions for fungal rot. Now's the time to reduce the risk of this disease by regularly going through your plants, picking off dead or yellowing leaves. Deadhead old flowers and space plants out as much as possible. Water sparingly and ventilate when temperatures allow. Clear up the plant debris from the greenhouse and put it on the compost heap.

In October you'll be right at the end of the tomato picking, and there may be a few chillies left to harvest, but over winter the late-sown salad leaves come into their own. If you have rocket, mustard, mizuna or cress, pick the outer leaves to keep them going through the cold months.

You can also try sowing cress and rocket now – they germinate at really low temperatures, as do peas for fresh sweet pea shoots in your salad. Pot-grown herbs will keep up the supply too. Top them up with pots from the supermarket.

JOBS TO DO THIS MONTH

- As natural light levels drop, let in as much light as possible by regularly clearing leaves and debris from the greenhouse roof.
- Clean pots, propagators and seed trays ready for using again in the spring.
- Insulate an unheated greenhouse with bubble wrap to ensure it is kept frost free.
- Sow seeds of hardy annuals for an early display next year.
- Clear out gutters to ensure a free flow of rainwater.

- Check brassicas and cut off yellowing leaves to help prevent fungal diseases.
- Prune any fading herbaceous perennials not being left for wildlife.
- Lift and divide perennials that have got too big.

Fresh from the Garden

This is a month where it doesn't feel like a great time to be on the veg patch. When it's not raining, it's blowing a gale, or everything's frozen. But that doesn't mean it's time to abandon the veg plot until next spring. You can still sow broad beans, plant rhubarb and get bare-root fruit bushes and trees into the ground. As well as this, there are still some harvests to be had. Some winter vegetables, including parsnips, kale and sprouts, actually get sweeter when it's frosty.

'Winter veg cope well with frost, but a really hard freeze can weld leeks, parsnips, carrots and beetroot into the ground, making them impossible to harvest. Tuck straw around your crops to prevent the soil freezing.' – **Sally Nex**

KEY TASKS

- Harvest parsnips.
- Plant rhubarb.
- Cover parsley with cloches.
- Prune blackcurrants, cutting the oldest shoots down to the ground.
- Cut down any remaining asparagus stems.

CROP OF THE MONTH: SWEDE

Did you know? – The swede may have got its name because it was sent as a gift to Scotland by the king of Sweden about 300 years ago. It's a tasty cross between a cabbage and a turnip.

Nutrients – High in vitamins C and E, fibre and calcium. Despite its sweet taste, it has only half the calories of a sweet potato.

Storing – Place in a paper bag if eating soon, or put in a shed or garage if storing for longer. Cool, damp conditions will help to prevent them shrivelling.

Good cultivars – For best flavour: 'Best of All' is mild with a smooth texture; 'Brora' is free from bitterness; 'Ruby' is extra sweet.

Sow – May–June

Harvest – September–December

HERB OF THE MONTH: OREGANO

The small, pungent leaves go well with chilli or garlic, and can be used fresh or dried. Oregano is a key ingredient in many southern Mediterranean dishes from pizza to pasta sauces and Spanish stews. Prefers a warm and sunny, sheltered spot. Cut this perennial herb back to the ground in winter or move indoors in autumn. This herb does die back in winter so if you want a harvest through the colder months, lift a plant, pot it and place in a light spot under cover.

GOT TEN MINUTES?

Sow broad beans for an early harvest

If you sow some broad beans now, they'll be ready to eat in May, a few weeks ahead of those that are sown next February or March. The seeds will germinate and the small plants will overwinter in all but the harshest winter conditions. Sow seeds 15cm apart in wide, 5cm-deep drills. Water them after sowing, then put supports in place for next spring.

GOT HALF AN HOUR?

Plant rhubarb

Rhubarb can be planted now. Add compost to the soil, water the crowns well and keep them at the soil surface when they are planted – they are likely to rot if completely buried. It's a good idea to allow the plants to strengthen for a year before picking the tasty young shoots the following spring.

GOT AN HOUR?

Plant a bare-root fruit tree

Most specialist nurseries supply young fruit trees lifted straight from the field once they are dormant. The lifting season lasts from November to February but the longer you leave it, the less choice of varieties you'll have. Most bush and tree fruits in this form will be ready to crop a year after planting. Prepare a weed-free site, add garden compost to the planting hole and drive in a stake for support. Use a mycorrhizal fungus on the roots before planting to improve establishment. Firm the soil around the roots, then water, mulch and guard against deer or rabbit damage in exposed gardens.

AND A FEW OTHER JOBS...

Bring in tender herbs

Bring herbs in now to keep a supply of fresh leaves growing. Put pots onto a bright windowsill or in the greenhouse. Tender herbs such as basil, coriander and dill won't last that long. Perennials such as mint and French tarragon will last for a while, but eventually need a dormant period so let them rest before the spring growth shows. Shrubby herbs can be kept in growth for much longer. Keep harvesting to boost productivity.

Harvest kale

Harvest kale leaves now. Start with the outer leaves and graze all your plants, taking just a few leaves from each. The plants will last all winter if you harvest like this. The most tender leaves are in the centre but if you take the whole crown there will be nothing left to crop except maybe a few spring shoots that will eventually emerge from the stump. The best strategy in future is growing enough plants to do both.

Rekha Mistry's Recipe of the Month

SAVOURY SQUASH ROLLS

Serves four

You'll need

250g winter squash or pumpkin, peeled, diced and steamed
360g strong bread flour
1½ tsp salt
¼ tsp turmeric powder
¼ tsp pav bhaji spice mix (or garam masala if not available)
¼ tsp dried coriander
7g fast-action dried yeast
2 tbsp rapeseed oil
Butter, for brushing the tops of the rolls
Sesame seeds, for sprinkling

Method

1 In a bowl, mash the steamed squash.
2 In another bowl, mix together the flour, salt, turmeric, pav bhaji spice mix, dried coriander, yeast and oil.

3 Stir the mashed squash into the flour and spice mix, adding a little water to bind it into a dough.

4 Knead the sticky dough for 10 minutes, then place in an oiled bowl to rise for about 1 hour.

5 Once the dough has risen, remove from the bowl and place on a floured surface – knead the dough for a few minutes. Make equal-sized balls and place on an oven tray, then let them prove for a further 30 minutes.

6 Bake the rolls at 190°C (fan) for 30 minutes. Remove from the oven and cool on a wire rack.

7 Brush the tops with melted butter and sprinkle sesame seeds over them. Serve with soup or curried veg.

Wildlife Notes

Things are quieter in the garden, now. There are few bees, except for the odd queen bumblebee on sunny days, and you may spot a late small tortoiseshell or red admiral butterfly feeding on the ivy flowers. Hoverflies, too, will gather here for the last of the year's nectar. Robins and blackbirds feast on rowan berries and windfall apples. If you plant more fruiting trees now – such as rowan, hawthorn, crab apple and eating apples – you will have more to offer the birds next autumn. Most trees have lost their leaves by now. Don't waste this resource – gather them into bags or a leafmould cage, and use to mulch your borders next year. Mulching at this time of year gives soil a boost and offers foraging for birds – the millipedes, beetles and grubs dwelling in the mulch providing a welcome snack.

LOOK OUT FOR...

- Fungi in lawns and on old logs.
- Goldcrests and firecrests – Britain's tiniest birds – darting between trees.
- Field mice, which don't hibernate.
- The last of this year's butterflies. Red admirals, peacocks and small tortoiseshells may be nectaring on late-flowering plants on sunny days.

- Dragonflies, some of which are still on the wing, resting around the pond edge or hunting for insects.
- Waxwings. Small reddish-brown birds with black throats, black masks round the eye and yellow and white in their wings, they may be looking for berries on cotoneaster or rowan.

'Look for the murmuration of starlings at dusk during autumn and winter. Our resident starlings are joined by visitors from Eastern Europe escaping the harsher winters. Just before roosting, they gather in huge, swirling masses to 'dance' into the sunset. This is thought to be a tactic to ward off predators – flying en masse, their synchronised movements look like vast, moving shapes in the sky.' – *Kate Bradbury*

3 WAYS TO HELP WILDLIFE IN WINTER

- **Add a hedgehog box** – If hedgehogs visit your garden, there's a good chance they will use a hedgehog box. Put it in the quietest part of the garden, ideally in shade, and fill it with hay or straw, for comfort. Surround it with logs so it looks natural.
- **Feed the birds** – Birds don't hibernate, so they have to find food throughout winter. If you don't have any berry-bearing shrubs or trees, now's the time to plant them – guelder rose, rowan, hawthorn and cotoneaster are ideal. Providing additional food is also essential. Choose peanuts, sunflower hearts and suet. The more calories the better, as these will give the birds the energy they need to survive cold nights.
- **Make habitat piles** – Pile up sticks, leaves and logs for insects, mammals and amphibians to shelter in. You can hide these at the back of the border or make a feature of them. Birds such as blackbirds, wrens and robins will pick through them for food. Avoid burning garden waste as you'll undoubtedly burn insects among the matter, too.

Spotter's Guide to Wintering Birds

In winter, our garden birds may look familiar, but many have travelled long distances to be here. Britain's largely ice-free winter climate means that they come here in autumn from Northern and Eastern Europe to take advantage of our mild weather.

Some, such as the redwing, fieldfare, brambling and waxwing, are almost exclusively winter visitors to Britain. Others, such as the robin, blackbird, blackcap, chaffinch and starling, are also here in spring and summer, but their numbers are augmented during autumn and winter by additional migrants that have flown from continental Europe; some from as far away as northern Russia.

All of these birds require our help: they rely on the food provided by us to get the energy they need to survive the winter, so they can return to the north to breed next spring.

5 BIRDS TO LOOK OUT FOR

- **Redwing** – The smallest British thrush: darker than the song thrush with a creamy stripe above the eye and red-tinged patch on the flanks. Redwings travel from Iceland and Scandinavia and feed largely in open countryside.
- **Chaffinch** – Vast numbers of chaffinches arrive in Britain each winter boosting the population of Britain's second commonest breeding bird. Males have a pink breast while females are buff in colour. Both show flashes of white in the wings when they fly.
- **Waxwing** – The ultimate prize for garden birdwatchers: this stunning bird arrives in flocks each autumn and is often found on berry bushes in gardens. Starling sized with pinkish-brown plumage, wispy crest and red on the wing – like sealing wax!
- **Brambling** – The northern relative of the familiar chaffinch arrives in variable numbers each year. Normally found in woods, it may also visit gardens to feed. Look out for the male's black head and wings, orange breast and bright white rump.
- **Robin** – Your familiar garden robin may be joined in autumn and winter by shyer relatives from the north and east. Sometimes this leads to fights between rivals as they each defend their patch of food.

 # Troubleshooting Guide

CLUB ROOT

This nasty infection causes root problems, meaning your plants fail to thrive. It only affects one family of plants – brassicas. Brassicas include cabbages, cauliflowers, Brussels sprouts, swedes and turnips, plus popular flowers such as wallflowers, and many weeds. The tiny organism that causes club root, a protozoa, lives in soil, where it can survive without a host plant for more than 20 years!

Roots swell and distort, so they don't work properly – the plants may wilt, go yellow, fall over or fail to grow. Proof of infection is obvious when you dig up your brassica vegetables at the end of the season. If you find it, take care to destroy the roots, firstly by drying and then by burning them. Club root is worse in acidic and damp soil, so liming and good drainage are advised if your soil is susceptible.

You should also take care with soil hygiene – don't move soil around and do disinfect boots, tools and wheelbarrows. Resistant cultivars are available, so buy these if you are troubled by the problem.

UNRIPE FIGS

In warmer countries, fig trees crop twice a year, but this is rare in the UK, unless they're under glass. By autumn, fig trees usually have two kinds of immature fruits. There are lots of hard green fruits of moderate size, which are destined to turn black and fall off. Near the tips you'll also see tiny pea-sized embryonic figs, which will ripen next summer. It's best to remove the doomed unripe figs in November. If you don't, they'll grow in mild spells, but their skins are so hard that they'll split and the fruits will drop off. The effort tends to weaken the plant, which may then fail to grow strongly enough to swell the pea-sized fruits to a mature size, especially in wetter, cloudier parts of the UK. As a result, these may stay green into autumn, and the plant gets locked into a cycle of unripe fruit.

POTTED TENDER PLANTS

Protect potted plants like bay trees, olives, bottle brush and myrtle by wrapping with horticultural fleece. This will trap a layer of air around them. A single layer can give an extra 2–3 degrees of frost protection, which can help protect perennials from the first frosts. For plants such as tender fuchsias and pelargoniums, it's best to bring them undercover to a greenhouse, conservatory or porch, where the temperature can be kept at 3–8°C.

DECEMBER

December weather can be the worst of the year in that it is grey, wet, cold and dank. However, if we get cold, frosty weather then the garden can be transformed. The mud is hard enough to walk on, the air is bright and fresh and the whole garden rimed and glittering with frost; suddenly all kinds of winter jobs become possible, from pruning to gathering crisp, semi-frozen leaves. Christmas decorations can often be gathered from the garden or the countryside, taking holly from the hedges, where the birds have not eaten all the berries.

'During Christmas week, I relish going outside to gather as much greenery as the garden will offer. A few berries from the holly will add a dash of colour, but I appreciate its unmatchable glossy greenness just as much. Holly recovers easily from an eager florist's hacking, its ability to submit to hard pruning making it perfect for topiary.' – **Monty Don**

 ## Weather Watch

A common question for weather forecasters this month is: 'Will it be a white Christmas?' Actually, snow is more common at Easter than at Christmas in most lowland parts of the British Isles. The surrounding seas hang on to small remnants of summer warmth, as water loses heat relatively slowly, and this can lend a little mildness to approaching weather systems.

Air frost is a different matter, and the long December nights allow plenty of time for temperatures to fall. However, those surrounding seas play their part here by reducing the incidence of air frost, which means that more tender plants can survive the winter closer to the coast.

WEATHER FACTS

- Highest regional rainfall: northern Scotland, 263.7mm
- Lowest regional rainfall: East Midlands, 49.3mm
- Average max temperature: 6.7°C
- Average min temperature: 1.1°C
- Highest number of days of air frost: East Anglia, 13.7
- Lowest number of days of air frost: western Scotland, 7.5

WEATHER ALERT

Frost in the garden damages tender plants, young foliage, flowers and even some forming fruits. If vulnerable plants are growing in pots, move them under cover or close to a sheltered spot, such as a south-facing wall. Wrap pots in bubble wrap and vulnerable plants in a double layer of fleece. Leaving last season's growth on tender plants will also provide some protection.

 Star Plants

10 OF THE BEST

1. Birch

The birch is one of Britain's best-loved trees. In the countryside we enjoy its white stems, its spring catkins and its buttery autumn colour, while in garden varieties we enjoy all those features with extra intensity plus its quick growth, light shade and dappled canopy of foliage. At this time of year the gleaming white bark of many varieties shines in the low winter sunshine. The weeping growth of some is elegant year round, while yellow spring catkins stand out against blue spring skies.

2. Holly

Nothing in the garden says festive like a big holly laden with bright berries. Through the dreariness of winter they sparkle like disco balls in a dingy nightclub. For something a bit different, varieties such as 'Argentea Marginata' have a subtle variegation that makes wreaths a bit special. If we go by the old carol – and on this occasion, with such a fabulous specimen, I really think we should – then, 'of all the trees that are in the wood, the holly bears the crown'. The runner-up, on that occasion, was the ivy, if you recall.

3. *Clematis cirrhosa*

If you thought that all the best clematises were around in the summer, then you are missing a bit of a trick. There are a few hardy, less flashy souls that do not come into their own until the darker days of December. This is an evergreen clematis with winter flowers. Varieties such as 'Freckles', which has cream flowers with red and brown speckles, or 'Wisley Cream' (creamy flowers) are really welcome at this time of year and perfect for a trellis, an obelisk or even an arch. Flowers December to February, depending on variety.

4. Italian arum 'Marmoratum'

As the clustered orange, berry-like fruits disappear (often taken by blackbirds), the foliage starts to appear. One glossy-green arrowhead follows another, from winter into spring, unfurling to reveal ivory ribs and veins. They seem impervious to even the most dreadful weather. The most majestic leaves, from the most mature tubers, have particularly pronounced markings and sensuous, undulating edges. Flowers June to August.

5. *Hippeastrum*

The flamboyant and oh-so-easy-to-grow windowsill *Hippeastrum* was once seen mainly as an emergency last-minute Christmas gift. *Hippeastrums* are available everywhere from garden centres to mail-order firms to garage minimarkets, but here's the thing – you may think that there are only two or three colours, but there were 68 varieties included in a

Royal Horticultural Society trial a
few years ago, including singles,
doubles, stripes and picotees,
ten of which received an Award
of Garden Merit. So, whether
you plant this month for flowering in spring, or you're already
enjoying flowers from bulbs planted
earlier, look beyond just the red and the
white. But take care as all parts of the plant are
toxic to cats. *Hippeastrum* bulbs can be forced,
so that they flower
in December.

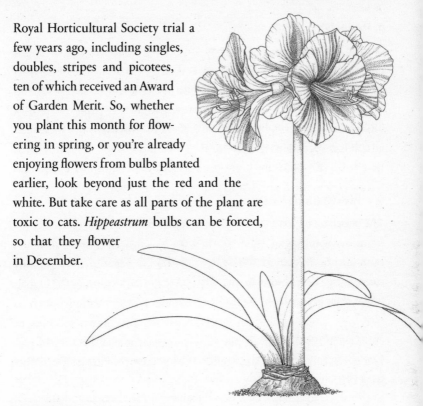

6. Mahonia

It's difficult to find many flowers around in December but mahonia
has scented sprays of yellow flowers during winter. It's also useful for
problem spots as it will thrive in poor soil and deep shade. *Mahonia
japonica* or 'Charity' are good varieties. Flowers November to March.

7. Cotoneaster

Cotoneasters such as *C. amoenus* are generally unappreciated: they are
mainly considered worth their place in the garden as background evergreens, small trees, hedges or as big green blobs grown to block off
unwelcome views. They are useful in all those roles but, at this time
of year, they come into their own. Clusters of Santa-coloured berries
scurry along every branch. When cut they make wonderful wreaths and
decorations; left in place they carry a frost with great aplomb and they
also provide valuable winter food for many of our garden birds.

8. Honesty

On the surface, the glory days of this lunaria are long past. Back in the spring, the mauvey flowers were scattered across our borders just as the sun began to warm our gardens. It was then forgotten among all the excitement of summer but now it is back and those little papery discs are catching the light and adding a bit of extra life to our winter. *Lunaria annua* 'Honesty' is the perfect dried flower for indoor arrangements. It's easy to grow, as it tends to self-seed. Flowers May to June.

9. Paperwhite narcissi

The scent of paperwhite narcissi is unmistakeable and will fill your home with fragrance this month. Like hyacinths and amaryllis, the bulbs can be 'forced' for festive flowers. 'Ziva' is one of the best narcissi for forcing. Keep bulbs in the dark for eight weeks after planting in bowls then gradually bring into the light. They will need support.

10. Wintersweet

Chimonanthus praecox is also known as wintersweet. It's the perfect name for a shrub with deliciously scented flowers that appear from December to February. Its small yellow flowers appear on bare stems and although they might not seem like much, they bring a sweet fragrance that will give your garden a lift in winter. Flowers December to February.

NOW'S THE TIME TO...

Make a wreath

Our gardens can offer up a surprising amount of greenery even in winter, and keep your eyes peeled when you're out for a walk – but don't gather plants from protected sites. Look for ingredients that will hold their shape as they dry out, so your wreath will last. You can also buy dried flowers, berries and seed heads from florist suppliers to add a sparkle to your decoration. By using foraged ingredients and avoiding plastic waste, you can celebrate the season in a planet-friendly way – plus, many of the elements can be reused next year, making it a thrifty celebration, too.

PLANTS TO ADD TO A WREATH

- Birch twigs
- Blue pine
- Douglas fir
- Viburnum with berries
- Rosemary
- Pine cones
- Honesty, dried
- Rosehips
- Ivy with berries

 Jobs to Do this Month

This a month not so much for gardening but on planning, tidying and decorating. The shortest day is approaching, there is little growth and low light levels. However, there are sheds to be tidied, catalogues to browse and decorations to make. It's also a key time to look out for wildlife and put out food for birds.

KEY TASKS

- Keep feeding the birds.
- Clean pots and tools.
- Bring in bulbs for indoor flowering.
- Move plants in pot to sheltered spots if the weather gets very cold.

GOT 15 MINUTES?

Target weeds

Target tough perennial weeds still going strong – they will now be much easier to spot and dig up. Remove as much of the roots as possible to help prevent regrowth.

GOT HALF AN HOUR?

Tend to winter containers

Tidy winter pots by deadheading and picking out windblown or damaged leaves. Look out for any slugs and snails that may be over-wintering beneath the foliage. Once you have tidied up, feel the compost to see if it needs some water. When it rains, the foliage gets wet but often the compost is left dry – so check this regularly, but allow it to dry out between each watering at this time of year when there is little growth. Raise the pot above the ground on feet or bricks to improve the drainage.

GOT AN HOUR?

Plant a bare-root shrub

Many woody plants, like fruit trees and shrubs, are supplied from the nursery with bare roots during the winter dormant period. Get them in the ground as soon as you can. First, soak the roots thoroughly, then plant in well-prepared ground. Look for the soil mark and plant to the same depth as it was at the nursery. If the ground is frozen or the site not quite ready, heel them into loose soil or compost, cover the roots and lay them down so that the branches are not vulnerable to the drying effects of the wind. You can delay planting for a few weeks like this, but they must be in the ground before the buds begin to swell.

AND A FEW OTHER JOBS...

Sort through your seeds

Get your seeds out of their storage tin and sort through them, ready for the sowing season. The harder the seed coat, the longer it will last, so make a judgement about whether to discard half-sown packets from last year. If you're really organised, you'll keep a list with dates for sowing them, and note what needs to be replaced and what to buy. Arrange the packets in groups of sowing times to make life easy when the spring rush starts.

Plant dogwood

You can plant rootballed or bare-root dogwoods between now and March; pot-grown kinds at any time – teasing out the roots if root-bound. Right after planting, shorten stems by a third to a half. To enjoy winter bark colour, cut established plants to within 10cm of the ground in March, as the youngest stems have the brightest colour. They'll give their best if never allowed to go short of water, so enriching soil with well-rotted compost or manure is beneficial. Once settled, they're not fussy and will thrive even in poor, dry earth.

Take root cuttings

December is the ideal time to take root cuttings of perennials such as mint, verbascum and primulas. Cut sections of root from the parent plant and divide into 5cm-long cuttings. Place in a tray filled with moist compost and cover with a thin layer of compost. Leave in a frost-free spot over winter.

'When it comes to usefulness, beauty and ease of growing, nothing beats the shrubby dogwoods with their colourful winter stems and (in the variegated kinds) brightly painted foliage. Planting them now [in December] makes sense, as they prefer damp soil and will establish quickly at this time of year.' – **Alan Titchmarsh**

ALAN'S TOP WINTER SHRUBS

- *Ilex* **'Golden King'** – Yellow variegated holly
- *Mahonia* **'Winter Sun'** – Yellow flowers, evergreen leaves
- *Sarcococca* – Evergreen with scented white blooms
- **Winter jasmine** – Masses of yellow flowers, great on walls
- **Witch hazel** – With orange, red or yellow blooms

DON'T FORGET TO...

- Move houseplants to a bright windowsill and turn every few days as the days become shorter.
- Clear dead leaves from the base of permanent container displays.
- Prune vigorous climbers such as Boston ivy, Virginia creeper and ornamental grapevines.
- Grit slippery paths.
- Turn the contents of compost bins to warm up on frosty mornings.
- Protect outdoor taps from freezing weather by taping bubble polythene over them.

LAST CHANCE TO...

- Bring in bulbs for flowering before Christmas.

 Time to Prune

Winter pruning needn't be difficult if you follow a few simple rules of thumb. The leafless branches mean you see what you're doing, so you can more easily choose what to keep and what to cut out. Just remember that winter pruning will stimulate subsequent shoot growth, whereas summer pruning tends to slow growth down.

Pruning now allows you to improve the shape of the plant, to remove overcrowded branches and any that make it lopsided. It can also encourage barren fruit trees to concentrate their energies on producing flowers and fruit, rather than a mass of vigorous leafy growth – provided you prune sensitively. If you just chop everything back to make a rounded dome, you'll end up with a forest of new growth the following spring. Selective pruning is the key.

PLANTS TO PRUNE

- Deciduous shrubs such as cotinus and berberis (not those that flower in spring/summer)

- Greenhouse grapevines
- Gooseberries
- Blackcurrants
- Climbing roses
- Deciduous trees
- Autumn-fruiting raspberries
- Apples and pears

In the Greenhouse

This month is a quiet one in the greenhouse but it's a good month to tidy up, clean and take on some tasks you might not have time for at other times of the year. Why not try taking root cuttings or sowing some onion seed instead of using sets. Check the temperature and ventilation in your greenhouse to keep plants healthy.

JOBS TO DO THIS MONTH

- Sow onions – growing from seed rather than sets gives more choice of varieties.
- Check your greenhouse temperature – some of your greenhouse plants may need some extra protection from the cold as the winter sets in.
- Mulch greenhouse borders – adding a layer of mulch to your greenhouse border now will condition the soil ready for next year's crop.
- Clean and oil your tools ready for the spring season.
- Stack your pots in size order to make sowing and potting easier this spring – once stacked, they'll also take up less room. Check them for cracks or splits.

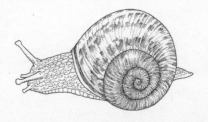

Fresh from the Garden

Although it is quiet on the veg plot, there can still be plenty to harvest with crops from collards, leeks, mustard greens and winter cabbages, along with bunches of herbs like parsley. Winter salads such as endive, rocket and 'Winter Density' lettuce will still grow in an unheated greenhouse. It's also the perfect time to start planning the new varieties of vegetables and herbs to add to the plot next year.

'There's one date in December that does excite me – 21 December. It's the shortest day in the calendar, when we know that slowly the daylight hours will start to increase. After this date, the heated mat comes out, seed trays get washed (again), seed packets are selected and seed compost bags tucked in the boiler room to warm up. And come 26 December, seed sowing can begin.' – **Rekha Mistry**

WHAT TO SOW

- Broad beans, under cloches
- Lamb's lettuce (indoors)
- Mustard greens (indoors)

WHAT TO PLANT

- Bare-root fruit trees and bushes

WHAT TO HARVEST

- Brussels sprouts
- Celeriac
- Celery
- Parsnips
- Turnips

CROP OF THE MONTH: CHICORY

Did you know? – All parts of chicory contain inulin, a soluble fibre that promotes gut health and can even help with weight loss.

Nutrients – Chicory is a good source of vitamins A, C and E, as well as calcium and iron.

Storing – Wrap in plastic to prevent drying. The chicons can keep for up to a week in a fridge.

Good cultivars – For best flavour: 'Totem' delicious and compact. 'Witloof Zoom' for strong taste.

Sow – May–June

Harvest – November–January

HERB OF THE MONTH: SAGE

Sage is a delicious and useful herb that can be picked year round and goes well with pork and chicken as well as many vegetable dishes. With its silvery evergreen leaves and pretty flowers – from intense blue to magenta – it looks good in the ornamental border and tastes good too. There are many different types of sage to choose from, so there's one to suit every situation.

Grow in well-drained soil in full sun. Annual and biennial sages can be grown from seed, while perennial sages are best grown from young plants. Many sages do well in pots. Harvest the leaves as and when you need to and trim back perennial types after flowering.

GOT 15 MINUTES?

Protect veg

Pull the earth up around the base of spring cabbages, broccoli and cauliflower with a hoe or trowel to provide protection from strong winds in winter. Remove yellow leaves to keep plants healthy and tie taller veg like Brussels sprouts to canes for extra support.

Harvest sprouts

Brussels sprouts taste sweeter when picked after some frost, so you can start harvesting this month. Simply snap each sprout cleanly from the stem by twisting it sharply downwards. Sprouts are always best eaten straight after picking, so only take as many as you need. The rest will stay tightly budded on the plant for some weeks to come. To prepare them, simply clean off the outer leaves and trim the base, then steam them gently. They will also keep for a few days in the salad drawer of the fridge, or can be frozen for use later.

Rejuvenate rhubarb

If you have a large, mature clump of rhubarb, you can dig it up now and divide it to rejuvenate the plant and produce several new ones. Lift the clump with as much root as possible, and cut off several sections of the youngest growth from the outside. Ensure each section has at least one strong shoot bud and lots of fleshy roots. Get your new rhubarb plants back into the ground as soon as possible. Add lots of organic matter to the soil, then position the plants so the shoot buds are at ground level. Water in well.

Inspect stored veg

Vegetables that you're storing, such as root crops, squash, garlic and onions, get progressively more vulnerable to rot as the weeks go on, even when kept somewhere cool and dark. So check them regularly – handle each one to make sure they're not going soft, and get some light on them so you can see if they've developed spots or other signs of rot on the skin. Onions, for example, can look perfect, but be soft when pressed. Use the softer ones as soon as possible, but discard any that have started to rot.

Put up bird feeders

Feed birds to keep them healthy over winter so they nest in your patch and help with pest management next season. Use a variety of feeders to provide seeds, nuts, grains and high-calorie foods such as fat balls. Place the feeders high up and away from any cover where predators like cats can wait to pounce. Keep the feeders clean and top them up regularly.

Plan next year's crop rotation

Take some notes or draw a sketch to record the positions where last season's crops grew. It's good to rotate crops, particularly the ones – like onions, potatoes and brassicas – that are vulnerable to soil-borne pests and diseases. Avoid growing any crop in the same place next year and think about the condition of your soil – root crops could follow potatoes since the ground has already been dug when lifting the tubers.

 # Rekha's Mistry's Recipe of the Month

PARSNIP KEDGEREE

Serves 2 as a main dish

You'll need

¼ tsp cumin seeds
2 tbsp sunflower oil
1 tsp green chilli, finely diced
1 tsp grated fresh ginger
1 small white onion, peeled and diced
1 tsp turmeric
1 tsp curry powder
Salt to taste
200g basmati rice, cooked
50g leftover roasted parsnips
50g leftover cooked Brussels sprouts and carrots (both optional)
Juice of ½ lemon

1 small red onion, peeled and sliced
2 eggs, boiled and quartered
¼ tsp chilli powder

Method

1 Heat a frying pan and toast cumin seeds prior to adding oil.
2 Toss in the diced chilli, ginger and white onion, and fry until the onion is soft, before stirring in the turmeric, curry powder and salt to taste. Then add a tablespoon of water to stop the spices from burning.
3 Add the rice, leftover vegetables and lemon juice, and stir until evenly mixed.
4 Garnish with red onion, arrange eggs and sprinkle over chilli powder, then allow to cook for a further 5 mins.
5 Serve with a cool carrot and mint raita.

Wildlife Notes

December consists of short days and long nights, which for creatures that do not hibernate means a scramble for food. The wildlife we are most likely to see in the garden during this month are birds, especially small species such as tits and wrens, which need to eat almost constantly throughout the day to give them the energy they need to stay warm at night. You may spot unusual visitors, too. Our smallest birds, the goldcrest and firecrest, often join roving flocks of tits to forage for food together. Look out for them pecking among outer tree branches for morsels such as moth eggs and spiders. On the ground, blackbirds, robins and wrens will be hungry, as their main source of food – worms and other invertebrates – are hibernating, and the ground may be frozen. Leave mixed seed, cooked rice and grated cheese for them, along with fresh water so they can drink and bathe.

DID YOU KNOW?

Tawny owls – pigeon-sized, brown and nocturnal – are more likely to be heard than seen. From dusk and through the night, their 'twit-twoo' call is actually two calls – the 'kee-wick' of the female, followed by the male's 'tu-woo'.

LOOK OUT FOR...

- Hedgehogs. If you see one out in the day, or evening, call a local hedgehog rescue for advice – it may not be fat enough to survive hibernation.
- Migrant birds (blackbirds or robins) may avoid bird feeders. Leave food and water out at the back of borders.
- Robins start pairing up and practising courtship rituals now. Look out for males feeding females – a sure sign of chicks in the coming weeks.
- Foxes don't hibernate, and some might start mating later in the month. Listen for females on heat shrieking.
- Bumblebees may emerge from hibernation on sunny days. If they seem unable to fly, mix a teaspoon of sugar with a teaspoon of warm water and pop it in an old bottle top, then place this next to the bee. She may drink it to give her the energy she needs to fly back to her hibernaculum.
- Honeybees, which emerge from the hive on sunny days and can get grounded if the sun disappears. Pop them on winter-flowering mahonia or honeysuckle for a sugary boost.
- Fieldfares and redwings may still be hunting for food in your garden.

Spotter's Guide to Ladybirds

Adult ladybirds overwinter in safe, dry corners, often in a small group. The first to arrive give off a 'safety' pheromone, attracting others to join. The chemical smell lingers until the following winter and the same sheltered nook may be used for years after by the same species. Harlequins, particularly, have a tendency to come indoors. These large, voracious invaders compete with, and even eat, native ladybird larvae, but after ten years here, our native species appear to be fine.

5 LADYBIRDS TO LOOK OUT FOR THIS MONTH

- **7 spot** – Large species, familiar to gardeners – red with seven black spots. They usually rest singly, in exposed places such as dead flowers, seed heads, the crook of a twig, a tightly curled dead leaf or the tied end of the washing line.
- **Harlequin** – Often gathers in corners or the inside of windows. Large, with varying patterns: straw-orange with 18–20 spots, to all black with 2 or 4 large red blotches. Always has two white blobs on thorax.

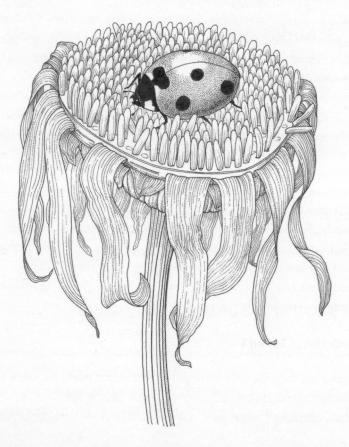

- **16 spot** – Tiny, beige to mustard-yellow with black spots, gathering in bunches of 10 to 30. Often rests on dead yellowed leaves, and is remarkably well camouflaged as the jumble of domes combine to form an irregular, stippled mass.
- **Pine** – This ladybird is black with four red spots (the front two spots look like commas). Usually rests singly, in a crevice where mortar has chipped from a wall, or a tight brick corner, or hard up against an ivy stem on a tree trunk. Overwinters on all types of tree, despite its common name.
- **2 spot** – Often seen in groups under loose bark on trees or logs. This ladybird is usually red with two black dots but it can show a bizarre range of spot patterns – the two black dots can become commas, semi-colons, or bars; even all black with four red marks.

 ## Troubleshooting Guide

MEALYBUGS

Mealybugs are likely to be found on quite a few houseplants around the country this month. These sap-sucking insects, shaped like mini-woodlice, can gather in multi-age-group colonies on all plant parts, but are often found in inaccessible places, like the gaps between leaves and stems. Here, protected from sprays and biocontrols, they are tricky to get rid of. If ignored, they soon multiply, producing white 'mealy' fluff, sticky sooty mould and discoloured patches on the plant, where they've sucked out the sap. A paintbrush or cotton bud will help you work insecticide into the place where they're lurking. The biocontrol ladybird can be introduced in spring, but not when you're using insecticides.

NO HOLLY BERRIES

If your tree lacks the jolly red or yellow berries that symbolise the festive season, the reason could be straightforward. Most hollies (*Ilex*) are dioecious, which means that male and female flowers are borne on separate plants. Some plants evolved like this to prevent self-fertilisation, thereby reducing inbreeding in a population and promoting the health of the species. But it can backfire when one of the sexes is vulnerable to change and starts to die out – no mate means no reproduction.

A tree that is male will have no berries, as only flowers with female parts can form fruits. Alternatively, your tree might be female but unpollinated. This happens when there is no male tree nearby with pollen-producing flowers. When you only have room for one holly, the answer is to plant 'J.C. van Tol', a tree that breaks the rules, being self-fertile and bisexual, producing red berries all on its own. You can plant several decorative females hollies like 'Lawsoniana' or 'Argentea Marginata' with just one male.

POINSETTIA PROBLEMS

A popular Christmas plant, the poinsettia has tiny flowers but it's the vibrant red, pink or white bracts that give it the wow factor. Prevent leaf damage after purchase by keeping it warm (12°C+). Even a cold

blast in a supermarket carpark can damage the foliage. Pick a spot at home with indirect light and only water it once the first few centimetres of compost are dry – overwatering will kill it! To keep them compact, poinsettias are often treated with growth retardant, which only lasts a season, while to keep them flowering and well coloured they need day-length control. Day-length sensitive plants, like poinsettia, only flower when they are exposed to the right amount of light. Poinsettias are short-day plants and produce floral buds instead of leaves when only lit for 12 hours.

As a result, some feel it's not worth trying to sustain them, but if you do want to, cut them back hard in early spring, grow them on, feed well, and create artificially equal day and night lengths from autumn by putting them in a cupboard 12 hours a day!

MISSING EARTHWORMS

If the soil is dry, a lack of earthworms may be because the worms are deeper in the soil awaiting moister conditions. But even if they have died, there may be eggs in the soil. As long as there are no acid-loving (ericaceous) plants in the border, give the soil a dressing of garden lime, as worms love this. Next, apply a mulch of well-rotted manure, leaf-mould or garden compost. Add a dressing of seaweed meal or blood, fish and bonemeal in spring. Keep the soil moist and regularly sprinkle a few grass clippings sparingly onto the surface, and the worms should return.

ACKNOWLEGDMENTS

We would like to thank everyone who has helped create this Almanac.

Many thanks to the team at *BBC Gardeners' World Magazine* and to their regular contributors, whose work has gone into this practical guide.

Special thanks go to Monty Don who has provided the inspiring foreword and to the Editor of *BBC Gardeners' World Magazine*, Lucy Hall, to Deputy Editor Kevin Smith and freelance gardening editor Tamsin Hope Thomson, who helped create this Almanac.

Our thanks to valuable contributions from Gardeners' World TV contributors, present and past, including Carol Klein, Alan Titchmarsh, Adam Frost, Frances Tophill, Arit Anderson, Mark Lane, Flo Headlam and Pippa Greenwood.

Magazine contributors who deserve a special mention include Rekha Mistry, for her seasonal recipes from the allotment; and Richard Jones for his Spotter's Guides; Kate Bradbury for her wildlife columns; Peter Gibbs for his weather statistics; herb guru Jekka McVicar, for sharing her expertise; and Sally Nex, for clear guidance on veg-growing.

For her elegant yet detailed illustrations, we are grateful to Christina Hart-Davies.

And for keeping the whole book on track, our thanks to project editor Jo Stenlake.

Weather stats based on 30-year average from 1981 to 2010. For more information on the weather stations supplying data, visit the Gardeners' World website

INDEX

Note: page numbers in **bold** refer to diagrams.

Abutilon 90, 198
  A. vitifolium 198
  A. x suntense 198
Acanthus 33, 174, 249
acclimatisation 70–1
Acer griseum 246
Achillea 118, 173
  A. 'Terracotta' 173
achochas 98
Aconitum 1
  A. 'Ivorine' 174
adventitious buds 78
African Lily *see Agapanthus*
Agapanthus 87, 224
agastache 249
Akebia quinata 112
Alchemilla mollis 146, 216
alfalfa sprouts 42–3, 64
algae 167, 204
Allium 110, 216
  A. fistulsum 132–3
  A. 'Purple Sensation' 110
  A. sphaerocephalon 172–3, **172**
Alocasia cucullata 224–5, **225**
alstroemeria 145
amaryllis 253, 288–90
Amelanchier lamarkii 54
amphibians 60, 74–5, 188
  see also frogs; newts; toads
anchusa 33

Anderson, Arit 122
Anemanthele lessoniana 226
Anemone
  A. nemorosa 137
  A. 'Pretty Lady Emily' 173
  A. 'Royal Blue' 55
  A. 'Vestal' 55
  Japanese 6, 59, 173
  spring 55
annuals 200
  cloches for hardy 271–2
  filling gaps with 151
  sowing hardy 37, 58, 151, 232, 274
  tender 155–6, 170
  see also specific annuals
ant hills 217
anthemis 117
aphids 57, 64, 87, 92, 99, 117, 129–30,
  133, 135, 182, 253, 257
apical dominance 103–4
apple 35, 37, 44, 62, 279
  'Arthur Turner' 113
  container-grown 132
  harvesting 220, 232
  non-fruiting 241–2
  planting 13, 38
  pruning 11, 180, 206, 273, 295
  storage 233, 258
  thinning out 183, 185
April 79–105

Aquilegia 6, 7, 110, 177
    A. 'Nora Barlow' 110
argyranthemum 229
artemisia 33
artichoke 93, 182
    globe 42, 161–2
    Jerusalem 39
        'Fuseau' 39
arum 'Marmoratum' 288
ash 22
asparagus 67–8, 93, 127, 275
    'Backlim' 127
    harvesting 93, 126, 130
    'Pacific 2000' 127
Aster
    A. 'Calliope' 245
    see also Symphyotrichum
Astilbe 174
    A. 'Professor van der Wielen' 174
Astrantia 7, 145, 174, 262
    A. 'Claret' 145
    A. major 'Alba' 145
aubergine 152–3, 155, 206, 232, 252
    'Listada de Gandia' 18
aubretia 82–3
August 195–217
autumn 220, 247
    colour 270–1
    see also November; October;
        September
autumn crocus 175
auxin 103–4
azalea 105

Bachelor buttons 84
banana plant 272
bark 251
basidiomycete fungi 216
basil 91, 115, 129, 155–6, 277
bats 134, 189, 214
    Daubenton's 188
    Pipistrelle 188
bay 15, 78, 206, 283

bears breeches *see Acanthus*
beauty berry 247
bedding plants 115, 151, 178–9, 248
bee hotels 100, 135–6, 163
beech 21–2
bees 19, 44, 55, 133, 154–5, 190, 213,
    279, 301
    bronze sweat 191
    cuckoo 191
    grounded 134
    hair-footed flower 73
    ivy 239
    leafcutter 134–5, 163, 164, 190,
        213
    red mason 100, 135, 190
    solitary 190–1
    white-faced 191
    wool carder 213
    see also bumblebees
beetles 279
    flea 186
    flower 134
    ground 104–5
    lily 166–7, **166**
    rosemary 104–5
beetroot 38, 182, 213, 262
    bolting 209
    container-grown 132
    harvesting 93, 153–4, 182, 208,
        254, 275
    leaves 130
    sowing 65, 125, 154
Begonia 139, 148, 150
    B. 'Apricot Shades' 121
berberis 294
bergamot *see Monarda*
bergenia 6, 139
bhajiya, kale 43–4
bidens 121
biennials viii, 182
bindweed 202–3, **203**
birch 99, 135, 163, 231, 287, 291
bird feeders 260, 281, 299

bird foot's trefoil, common 136
bird migration 259, 281–2, 301
bird pests 191
birdsong 135
black spot 192–3, 217, 251
blackberry 13, 77, 233
  harvesting 207, 220
  pruning 158, 252, 273
blackbirds 44, 72, 99, 163–4, 259, 279, 281, 300–1
blackcap 281
blackcurrant 11, 156, 182, 207, 252, 275, 295
blackfly 129–30
bleeding heart 84
blood, fish and bonemeal 40, 47–8, 61, 121, 305
blossom 112–13
blossom end rot 152, 181
Bluebeard 221–2
bluebell 137
blueberry 65, 114, 132, 156, 207, 220, 254
bluebottle 20
bolting 185–6, 209–10
bonfires 101
borders 249, 251
Boston ivy 294
botrytis 253
bottle brush 283
bougainvillea 37
box *see Buxus*
Bradbury, Kate 99, 280
bramble 48
bramblings 281, 282
brassicas 42, 183, 189, 257, 275
  crop rotation 299
  earthing up 183, 236
  feeding 187
  harvesting 234–5
  planting 182, 257
  problems 167, 233, 282–3
  see also specific brassicas

broad bean 16, 42, 64, 92, 126
  'Aquadulce Claudia' 16, 18, 237
  broad bean tortilla 162
  'Bunyard's Exhibition' 16
  'De Monica' 16
  early 277
  harvesting 126, 130, 142, 153
  leaves 126
  'Masterpiece Green Longpod' 16
  problems 129–30
  sowing 38, 65, 254, 275, 277, 296
  'The Sutton' 16
broccoli 42, 297
  Calabrese 211, 235
  harvesting 211, 234–5
  'Purple Sprouting Early' 94
  purple sprouting viii, 38, 94
  'Red Arrow' 94
  sprouting 64, 235, 257
Brunnera macrophylla 84
  B. m. 'Jack Frost' 84
Brussels sprouts 42, 297, 299–300
  harvesting 13, 38, 275, 296, 298
  problems 167, 282
  sowing 65, 93
Buddleja 188, 262
  B. davidii 63
  pruning 11, 62–3, 89, 251–2
bulbs 22
  autumn 175, 179, 224
  dwarf 55
  in grassed areas 59, 227–8
  planting depths 262
  spring viii, 23, 227, 228
  summer 80, 85, 87, 125
bumblebees 19–20, 44, 72–3, 99, 102, 163, 213, 279, 301
  buff tailed 102
  common carder 102, 238, 239, 259
  feeders 73
  red tailed 102

tree 102
white tailed 102
busy Lizzy, New Guinea 139
butterflies 133, 188
    brimstone 73
    comma 134, 188, 238
    feeders 46
    holly blue 215
    orange-tip 134
    painted lady 213
    peacock 72, 99, 164, 188–9, 279
    red admiral 20, 45, **45**, 188–9, 238,
        259, 279
    small tortoiseshell 189, 238, 279
    tortoiseshell 72, 99, 188
    yellow brimstone 45
Buxus 86, 201, 204–5, 216

cabbage 92, 262
    autumn 233, 254
    Chinese 234
    earthing up 236, 297
    harvesting 65, 211, 233, 234–5,
        296
    ornamental 248
    planting 126, 233, 254, 257
    problems 167, 282
    sowing 93, 97–8, 182, 208,
        234
    spring 182, 208, 211, 233, 254,
        297
    summer 97–8, 182, 211
    winter 296
cacti 152
cake, rhubarb and vanilla 98–9
calcium deficiency 181
Calendula 58, 88, 92, 151, 185, 271
calibrachoa 85
Callicarpa 247
callistemon 123
camassia 227
camellia 23, 26, 28–9, 49, 90
campanula 88, 118, 251

campion, bladder 137
candytuft 82–3
Canna 87, 153, 170, 172, 198, 272
    C. 'Phasion' 198
    C. 'Wyoming' 174
    overwintering 266
cardoon 268, **268**
carp, grass 77
carrot 96, 127–8, 182
    Chantenay Red Cored 95
    container-grown 132
    Early Nantes 95
    Flyaway 95
    harvesting 153, 183, 254, 275
    Parmex 95
    sowing 13, 95, 126, 127
    Sweet Candle 95
carrot fly 128
Caryopteris x clandonensis 'Blue
    Knight' 221–2
caterpillars 99, 133, 163, 167, 189,
    192, 214–15, 240–1, 257
catkins 1, 54, 287
catmint *see Nepeta*
cauliflower 65, 182, 297
    harvesting 234–5
    problems 167, 282
    sowing 125, 127
cavolo nero 236
celeriac 14, 158, 187, 296
    'Giant Prague' 14
    'Monarch' 14
    'Prinz' 14
celery 127, 158, 296
centipedes 47
centranthus 177
Ceratophyllum demersum 167
Ceratostigma 223
    C. *plumbaginoides* 223
    C. *willmottianum* 223
cerinthe 151
chaenomeles 53
chaffinch 281, 282

chamomile 185
chard 126, 209, 211, 254, 262
    Swiss 13, 65
checkerberry 269
cheesecake, a twist on raspberry
    258–9
Chelsea chop 117–18
cherry 78
    Cornelian 28
    harvesting 154
    Morello 77
    ornamental 112
    'Pink Shell' 112
chervil 94, 155
chickweed 138, 154
chicory 126, 154, 182, 297
    'Totem' 297
    'Witloof Zoom' 297
chillies 97, 127, 276
    chilli jam 237–8
    container-grown 132
    hardening off 129
    harvesting 183, 206, 208, 232–3,
        255, 274
    pickled 235
    planting 126, 153
    sowing 13, 38
Chimonanthus praecox 290
Chinese silver grass 223
chionodoxa 52, 227
chives 115, 140, 154, 185, 234
chlorophyll 76–7
chocolate vine 112
Choisya 111
    C. 'Aztec Pearl' 111
Christmas rose 3
Christmas trees 8–9
chrysanthemum 216, 229, 245,
    262
cistus 123
citrus 252–3
Clematis 145, 153
    boundary cover with 121

C. *alpina* viii
    C. *a.* 'Frances Rivis' 30, 111
C. *armandii* 55
    C. *a.* 'Apple Blossom' 30, 49
C. *cirrhosa* 288
    C. *c.* 'Freckles' 288
    C. *c.* 'Wisley Cream' 30, 288
C. 'Étoile Rose' 122
C. *montana* viii, 121
    C. *m.* 'Elizabeth' 30
    C. *m. rubens* 122
C. 'Nubia' 200
C. *orientalis* 121
C. 'Perle d'Azur' 122
C. 'Sieboldiana' 145
C. *tangutica* 121, 122
C. *texensis* 121
C. *viticella* viii, 121
    C. *v.* 'Purpurea Plena Elegans'
        122
C. *x cartmanii* 'Avalanche' 30
container-grown 121
cuttings 177
early-flowering 30, 111
late-flowering viii, 35, 62
planting tips 121
winter 19, 20
climbers 49, 55, 86, 120, **120**, 294
    see also specific climbers
cloches 27, 271–2, 275
club root 282–3
cobaea 88
colchicum 179
cold frames 71, 92, 129
collards 13, 296
comfrey tea 125, 154
compost 8–9, 42, 61, 122, 131, 140,
    158–61, 230, 251, 305
compost bins 74
compost heaps 85
containers 88
    feeding 131, 149, 165–6, 181
    fruit in 95, 132

hanging baskets 84–5, 147–9
salads in 96–7
for shade 105
summer 119–20, **119–20**, 225–6,
226
top dressing 34
top plants for 121, 200
vegetables in 131–3
watering 88–9, 120, **120**, 131, 151,
165, 171, 178
winter 5–6, 34–5, 292
coriander 91, 115, 252, 277
'Leisure' 71
corms 85, 113
corn salad 208, 254
cornflower 85, 151, 183
Cornus 4, 6, 11, 62, 269, 293
C. controversa 'Variegata' 246
C. mas 28
C. sericea 'Flaviramea' 6
Cosmos 88, 173
C. 'Sonata White' 173
Cotinus 63, 247, 294
C. 'Grace' 271
Cotoneaster 20, 280, 281, 289
C. amoenus 289
courgette 37, 92, 208–9
'Black Forest' 209
blossom end rot 181
container-grown 132
feeding 187
flowers 85
'Gold Mine' 209
hardening off 129
harvesting 183, 185, 207–8, 211,
232–3
sowing 93
'Venus' 209
cowslip 54, 136
crab apple 112, 279
Japanese 112
creeping Jenny 85
cress 274

crimson flag *see Hesperantha*
crinum 179
Crocosmia 197
C. 'Emily McKenzie' 197
C. 'Lucifer' 197
crocus 26, 52, 72, 99, 227–8
crop rotation 92, 299
Crown imperial 262
cucumber 91–2, 158, 181
container-grown 132
harvesting 206, 207, 252
planting out 155
sowing 64, 124
currant
flowering 151, 152
red 77, 156, 182, 186, 207
white 77, 156
see also blackcurrant
cut flowers 175
cuttings 149–50, 185
root 33, 185, 293
softwood 115, 176–7
stem 118, 185
tender plants 115, 229
Cyclamen 179
C. coum 4
C. hederifolium 216

daffodil 26, 44, 52, 55, 57, 227–8
Dahlia 22, 151, 170, 172, 182, 227,
245
cactus 198–9
cuttings 30
D. 'Biddenham Strawberry'
31
D. 'Bora Bora' 31
D. 'Chat Noir' 199
D. 'David Howard' 31, 172
D. 'Gay Princess' 31
D. 'Kenora Valentine' 31
D. 'Magenta Star' 200
D. 'Primrose Diane' 31
deadheading 175, 179

overwintering 248, 266, 271
  planting out 30–1, 85, 87
damselfly 134, 165, 189
dandelion 48, 103
Daphne 34, 112
  D. cneorum 112
  D. laureola 217
  D. mezereum 28
daylily 59, 249
deadheading 147–8, 152, 175, 178–9,
  191–2, 200, 203–4, 274
deadnettle, red 137
December 285–305
deer 48
delphinium 88
deutzia 151, 180
Dicentra 84
Digitalis viii, 6, 144, 152, 177, 182
  Canary Island 199
  D. purpurea albiflora 216
dill 91, 115, 127–8, **128**, 277
  'Bouquet' 128
  'Diana' 128
division 59, 87–8, 185, 221, 275
dogwood *see Cornus*
Don, Monty vii–ix, 15, 30, 62, 88,
  146, 159–60, 178, 210, 286
Doronicum orientale 55
Douglas fir 291
dragonflies 165, 189, **189**, 280
  broad-bodied chaser 133
  red darter 164
duckweed 77
dunnock 21

earwigs 47, 227
echinacea 33, 117–18, 223, 249
edible flowers 85
elephant ear 224–5, **225**
endive 296
Epimedium 55, 82, 105
  E. 'Fire Dragon' 82
  E. x perralchium 'Frohnleiten' 82

equinoxes
  autumnal 139, 220
  vernal 52–3, 139
Equisetum arvense 139
Eremurus 271
Erica carnea 'Vivellii' 4
Erigeron karvinskianus 144
Erysimum 'Bowles Mauve' 29
Erythronium 55, 82, 146
  E. 'Citronella' 82
escallonia 123
Euonymus 152, 269
  E. alatus 267
  E. fortunei 269
    E. f. 'Emerald Gaiety' 217
  E. japonicus 269
euphorbia 26
European monsoon 142–3
evergreens 22, 28–9, 38, 83, 111,
  273
  climbers 49, 55
  leaf miners 262
  moving 230
  pruning 63, 150, 152
  see also specific evergreens

fat hen 154
February 25–50
feeding plants 40, 47–8, 61, 89, 154
  Clematis 121
  containers 131, 149, 165–6, 181
  DIY organic feed 125
  in greenhouses 152
  lawns 76–7, 85, 89
  sign you need to 165–6
  summer crops 187
fences 49, 77
fennel 42, 207, 209, 222, 254
fern 90, 146
  Boston 146–7
  Japanese lace 34
  soft field 28
fertiliser *see* feeding plants

fieldfare 20, 259, 281, 301
fig 132, 283
finches 188
firecrest 279, 300
fish 74–5, 77
fleabane, Mexican 144
fledglings 163, 164
fleece 9, 22–3, 42, 61, 117, 125, 162, 262, 272, 283, 286
flies 20
flocculation 75
flowering period extension 117–18
Foehn effect 2
Fontinalis antipyretica 167
forget-me-not 182
forsythia 55, 63, 151, 153
fothergilla 105
fox 19, 188, 301
foxglove viii, 6, 144, 152, 177, 182, 216
    Canary Island 199
foxtail lily 271
French bean 88, 126
    container-grown 132
    harvesting 208
    sowing 127, 155, 157, 182
fritillary 166
frogs 74–5, 77, 99–100, 105, 163, 189, 238, 259
    common 74
frogspawn 44, 72, 73
frost 22–3, 26, 28, 31, 42, 92–3, 108–9, 151, 162, 170, 245, 252, 259, 266, 271–2, 275, 283, 286–7
Frost, Adam 16, 138, 177, 224, 256
fruit
    and bird pests 158, 191
    harvesting 183
    for north-facing fences 77
    storage 258
    top dressing 95
    see also growing your own fruit and veg; *specific fruit*

fruit bushes 233
    mulching 93
    planting bare-root 275, 296
    protection for 208
    pruning 11, 180
    see also specific types
fruit trees 17, 155, 233, 279
    and pests 257
    planting bare-root 277, 296
    pruning 180, 273
    see also specific types
fuchsia 37, 62, 148, 216, 283
    Australian 34
    cuttings 150, 177, 229
    hardy 35, 90
fungal diseases 181, 192–3, 206, 216–17, 232, 251, 253, 274, 275, 282–3
fungi
    basidiomycete 216
    mycorrhizal 76, 277

galtonia 61
garland flower 112
garlic 276, 298
    elephant 179
    'Germidour' 256
    harvesting 153, 183, 186
    'Picardy Wight' 236
    planting 254, 255, 256
    'Solent Wight' 256
Gaultheria procumbens 269
Geranium 6, 37, 111, 148, 174, 229, 252
    cranesbill 111
    G. phaeum 'Lily Lovell' 216
    G. pratense 174
    G. robertianum 136
    G. 'Rozanne' 111, 200
    G. 'Samobor' 216
Geum 110
    'Mrs J Bradshaw' 110
ginger 170
Gladiolus 61, 87, 113, 175
    G. callianthus 113

G. 'Espresso' 113
G. 'Green Star' 113
G. 'Plum Tart' 113
gluts 211
goldcrest 20, 279, 300
goldfinch 259
gooseberry 13, 77, 155–6, 182
    'Invicta' AGM 155
    'Rokula' 155
    training 11, 186, 273, 295
    'Xenia' 155
grafted plants 78
grape hyacinth *see Muscari*
grapevine 11, 152, 273, 294–5
grasses 63, 90, 99
    see also lawns
grasshoppers 189, 214
greasebands 257
greenfly 64, 92, 117
greenhouses 9, 64, 152–3, 231–2, 274–5, 283
    in autumn 252–3
    cleaning 12, 64, 232, 253, 274
    cold 24
    cooling 153
    heating 12, 24, 124, 253, 274
    insulation 274
    problems 253
    shading 91–2, 124, 152–3, 181, 206
    in spring 91–2, 109, 124–5
    in summer 181–2, 206–7
    temperatures 274, 295
    ventilation 64, 91, 109, 153, 181, 206, 232, 274
    in winter 12, 37, 295
growing your own fruit and veg 64–71, 153–62, 232–7, 275–8
    autumn 254–8
    clearing old crops 257
    for small spaces 140
    spring 92–8, 125–33, 140

summer 182–7, 207–12
winter 13–18, 37, 38–43, 296–9
guelder rose 271, 281

habitat degradation 190
half-hardy plants 37, 123, 198
Hamamelis 26, 293
    H. x intermedia 'Pallida' 3
hanging baskets 84–5, 147–9
hardening off 92, 117, 129
hawthorn 99, 112, 279, 281
    double pink 113
hazel 78, 99
Headlam, Flo 225–6
heather 4, 248, 272
    winter 4, 85, 90
Hebe 63, 204, 246
    H. 'Caledonia' 246
    H. 'Claret Crush' 6
    H. 'Mrs Winder' 246
hedgehog 44–5, 99, 100–1, 133, 135, 188, 238, 259, 281, 301
hedges
    bare-root 271
    planting 86–7, 271
    pruning 201, 204–5, 231, 251
Helenium 59, 117, 223, 251
    H. autumnale 'Salsa' 226
Heliotropium arborescens 200
Helleborus 1, 6, 9, 26, 34, 48, 55, 105, 139, 262, 272
    H. niger 3
herb Robert 136
herbicides, glyphosate 203
herbs 42, 80, 114–15, 127, 153
    hardening off 129
    harvesting 130, 211
    pot-grown 132, 274
    sowing 91
    tender 252–3, 277–8
    see also specific herbs
Hesperantha 247

Heuchera 5, 26, 105, 139
  'Midnight Rose' 5
hibiscus 123
Hippeastrum 253, 288–90
hoeing 67, 138
Holboellia brachyandra 49
holly *see Ilex*
hollyhock 216
honesty *see Lunaria*
honeysuckle *see Lonicera*
hornbeam 22
hornets 190
hornwort 167
horse chestnut 262
horsetail, field 139
hosta 59, 249
houseplants 146–7, 294
hoverflies 73, 87, 182, 238
  marmalade 44
hyacinth 29, 228, 290
  see also Muscari
Hydrangea 57, 62, 90–1
  H. paniculata 35, 62
  H. quercifolia 271
Hylotelephium (sedum) 59, 90, 118, 222–3
Hypericum 248
  H. calycinum 90

Iberis sempervirens 82–3
  I. s. 'Snowflake' 83
Ilex 44, 86, 152, 262, 286, 288, 303–4
  I. 'Argentea Marginata' 288, 304
  I. 'Golden King' 293
  I. 'J.C. van Tol' 304
  I. 'Lawsoniana' 304
insecticides 101, 192, 253, 262
International Dawn Chorus Day 135
ipomoea 88
Iris 1, 26, 52
  bearded 111, 176

'Black Swan' 111
early miniature 27
I. histriodes 'George' 27
I. reticulata 'Katherine Hodgkin' 27
ivy 4, 44, 239, 260, 279, 288, 291

jam, chilli 237–8
January 1–24
Japanese squash 'Uchiki Kuri' 132
jasmine
  common 48, 174
  star 49
  summer-flowering 252
  winter 293
Judas tree 112
July 169–93
June 141–68

kale 92, 93, 127, 182, 257
  harvesting 13, 38, 130, 211, 232–4, 275, 278
  kale bhajiya 43–4
  for small spaces 140
katsura tree 271
kedgeree, parsnip 299–300
Kerria japonica 'Pleniflora' 84
kestrel 134
kiwi 273
Klein, Carol 6, 56–7, 112, 174, 200, 230, 270–1
knautia 177
Kniphofia 59, 199, 223
knotweed, lesser 146
kohlrabi 154, 186, 213

lacewings 257
ladybirds 87, 99, 135, 182, 192, 214, 257, 301–2, **302**
  2 spot 303
  7 spot 302
  16 spot 303
  harlequin 239, 301–2

pine 303
seven-spot 239
lamb's lettuce 157, 296
Lamium purpureum 137
Lamprocapnos spectabilis 84
Lane, Mark 252
lantana 151
Lavandula 7, 85, 104, 172, 201
    L. 'Arctic Snow' 172
    pruning 85, 203–4, 206, 251
lavatera 35, 262
lawns 139–40, 250
    ant hills on 217
    bare patches on 167–8
    brown 241
    bulbs in 59, 227–8
    edge repair 60
    feeding 76–7, 85, 89
    mowing 57, 59, 77, 89, 203
    patchy 76–7, 207
    scarifying 139, 207, 227, 241,
      250
    sowing 167–8, 228
    wormcasts on 22–3, 58
leaf discolouration 165
leaf miners 133, 262
leafmould 266, 272–3, 279, 305
leek 38, 42, 110, 127, 187
    harvesting 14, 38, 275, 296
    planting 126, 154, 182
    red lentil and leek soup 71–2
    sowing 13, 38
lemon 132
lemon thyme 256
lemon verbena 151
lenten rose 34
lentil, red lentil and leek soup 71–2
leopard's bane 55
lettuce 42, 49, 108, 156–7
    butterhead 211
    container-grown 96–7
    cos 211
    cut-and-come-again 211

    'Green Oakleaf' 96–7
    harvesting 126, 130, 153, 207, 211,
      254, 296
    Iceberg 17
    'Lollo Rossa' 157
    Mesclun mix 157
    'Red Salad Bowl' 96–7
    'Salad Bowl' 96–7, 157
    sowing 64, 65, 92, 156–7
    winter 13, 38, 236–7, 296
    'Winter Density' 236–7, 296
Leycesteria 35
Libertia ixioides 'Taupo Sunset' 226
lilac *see Syringa*
lily 61, 153, 166, 175
    daylily 59, 249
    potting up bulbs 60–1, **60**
    Turk's cap 145
lily of the valley 112
lime 14, 139, 305
Liquidambar styraciflua 269
lobelia 85, 150
log piles 101, 281
Lonicera 123, 174, 231, 301
    L. fragrantissima 29
    L. x purpusii 29
    winter 19, 20, 29
Lotus corniculatus 136
Lunaria 291
    L. annua viii, 167, 290–1
lungwort *see Pulmonaria*
lychnis 177

Magnolia wilsonii 110
Mahonia 19, 20, 105, 289, 301
    M. 'Charity' 289
    M. japonica 289
    M. 'Winter Sun' 293
Malabar spinach 187
    Malabar spinach sauce 187–8
manure 40, 74, 156, 305
    chicken 40, 47–8, 131, 156
    green 254, 255

maple, paperbark 246
March 51–78
marigold *see Calendula; Tagetes*
marjoram 42
marrow 208
martins, house 134, 163, 188, 189, 213
May 107–40
mealybugs 303
melon 124, 152, 185–6, 232
mibuna 254
mice 32, 99
    field 19, 20, 189, 279
micro greens 64
migration 259, 281–2, 301
mildew 192–3
milfoil, spiked water 167
millipedes 279
    pill 47
    snake 47
mint 115, 184–5, 278
Miscanthus sinensis 'Red Cloud' 223
Mistry, Rekha 18–19, 43–4, 71–2, 98–9, 129, 162, 183, 187–8, 207, 212–13, 237–8, 258–9, 278–9, 296, 299–300
mites, red spider 37, 124
mizuna 234, 254, 274
moles 193
Monarda 173, 249
    M. 'Gardenview Scarlet' 173
    M. 'Scorpion' 173
monkshood *see Aconitum*
Monstera 224
mooli (daikon radish) 17
    mooli wraps 18–19
mosquitos 216
moths 99, 133, 214, 257
    adela 4134
    cinnabar 215
    codling 240–1
    drinker 215
    elephant hawk 215

garden tiger 164, 214
six-spot burnet 214
mulches 40, 61, 87, 162, 279
    for borders 251
    for clay soils 75
    compost 267
    and earthworms 305
    and fungal diseases 192, 217
    and water retention 81, 153
    and weeds 138, 139, 153
Muscari 52, 81–2, 99, 228
mustard greens 234, 274, 296
mycorrhizal fungi 76, 277
Myriophyllum spicatum 167
myrtle 283

Narcissus
    N. 'February Gold' 29
    N. 'Rijnveld's Early Sensation' 29
    paperwhite narcissi 290, 291
nasturtium 58, 167, 213
naturalising 227
nematodes 86, 89, 182, 217, 240
Nepeta 172
    N. *racemosa* 'Walker's Low' 172
Nerine 179, 224
    N. *bowdenii* 224
netting 158, 167, 191, 208
nettle 99, 122–3, 164
    see also deadnettle
newts 73–5, 100, 188–9, 238, 259
    smooth (common) 74–5, 164
Nex, Sally 70, 211, 275
nigella 151, 232, 271
nitrogen 131, 158
North-facing plots 139, 216
Notonecta 188
November 265–83

oak 99
    English 21
October 243–63
olearia 123

olive 262, 283
onion 42, 70, 110, 158
  autumn-planted 42
  bolting 209
  crop rotation 299
  harvesting 183, 208
  Japanese 233
  planting 65, 93, 254
  sowing 13, 233, 295
  storage 298
orchid 152
oregano 256, 276
organic matter 75, 81, 267
Osmanthus delavayi 112
owls, tawny 301

pak choi 154, 234
pansy 248
parsley 66, 140, 155, 206, 275, 296
  flat-leaved 252
parsnip 14, 127, 275, 296
  parsnip kedgeree 299–300
pasque flower 7
passion flower 37
pea 42, 182
  harvesting 142, 153, 183, 185, 211
  sowing 38, 65, 92–3, 125, 127, 254
pear 35, 37–8, 62, 77, 216, 258
  harvesting 220, 232
  pruning 11, 180, 206, 273, 295
pelargonium 121, 149–50, 177, 206,
  216, 229, 241, 252–3, 283
Pennisetum 268
  P. 'Rubrum' 268
Penstemon 118, 222
  cuttings 118, 177
  P. 'Schoenholzeri' 222
  P. 'Sour Grapes' 200, 222
  pruning 62–3, 85, 123
peony 143
pepper 127, 213
  harvesting 206, 207, 232, 233, 252
  planting 126, 153, 155

perennials
  bare-root 34
  and the Chelsea chop 117–18
  division 59, 275
  moving 59
  planting 230, 247
  pruning 231, 275
  sowing from seed 6–7
  tidying up 180
  trimming back 33–4, 61
  see also specific perennials
Persicaria 246
  P. affinis 146
pesticides 101
pests
  and brassicas 167, 233
  and container veg 131
  and fruit trees 257
  hunting 257
  see also specific pests
petunia 85, 148, 150
pheasant's tail grass 226
philadelphus 151
Phlox 118, 171–2
  night 174
  P. paniculata 'Blue Paradise' 171–2
Phormium tenax 6
photosynthesis 76–7
phygelius 62, 63
pickling 235
pinching out 117, 126, 153, 186
pine 291
pinks 175
Pittosporum 123, 269–70
  P. tenuifolium 270
plug plants 115–17
plum 13, 78, 208
plumbago see Ceratostigma
poinsettia 304–5, **304**
polyanthus 54, 105, 248
polypody, common 34
Polystichum setiferum 28
pond plants 87–8, 177–8

ponds 10, 60, 100, 164–5, 189
  cleaning 77, 167, 204
  topping up 201
pondskaters 164–5
poplar 78
poppy 48
  'Ladybird' 197
potassium 131
potato 19
  chitting 40–1
  crop rotation 299
  'Desiree' 125
  Duke of York 41
  earlies 41, 57, 154
  earthing up 126
  Foremost 41
  harvesting 153–4, 185, 207, 210,
    236, 254
  International Kidney 41
  'King Edward' 125
  Pentland Javelin 41
  planting 93, 97, 125, 126
  Rocket 41
pricking out 104, 118, 124, 127
primrose 53–6, 85, 99
Primula 26, 53–4, **54**, 56–7, 105
  P. bulleyana 56
  P. Gold-laced Group 56–7
  P. veris 54, 136
  P. vialii 56, 146
  P. vulgaris 54, 56
    P. v. 'Lilacina Plena' 56
privet 78, 201, 204–5
propagation
  pond plants 87–8
  rosemary 39
  see also cuttings; division; seeds
propagators 30–1, 42
protecting plants
  from birds 158
  from winter cold 3, 9, 22–3, 27,
    42, 61, 70–1, 93, 125, 271–2,
    297

pruning 35–6, 89–91, 151–2, 231,
  273–4
  autumnal 251–2
  climbers 294
  lavender 203
  perennials 231, 275
  plum 208
  spring 62, 123
  spur 205
  summer 180, 204–6
  trimming hedges 201
  viburnum 49–50
  winter 10–11, 12, 294–5
  see also specific species, pruning
Prunus angustifolia 54
Pulmonaria 26, 54–6, 105
  P. 'Blue Ensign' 54, 56
pumpkin
  'Amazonka' 255
  bird feeders 260
  harvesting 236, 254, 255
  'Jack Be Little' 255
  'Pot of Gold' 255
  savoury squash rolls 278–9
purple bell vine 222
pussy willow 54
Pyracantha 20, 44, 204, 206–7, 270

quince 38

radish 42, 64, 183–4, **184**, 213
  'Black Spanish' 184
  'French Breakfast' 184
  harvesting 93, 126, 154
  sowing 13, 92, 154, 208
  'Sparkler' 184
ragwort 215
rainfall 3, 27, 53, 80–1, 109, 142–3,
  171, 196–7, 221, 245, 266–7, 286
rainwater storage 197
raspberry 38, 114, 154, 182
  autumn 233–4, 295
  'Autumn Bliss' 234

harvesting 156, 207, 220, 233
'Himbo Top' 234
'Polka' 234
pruning 12, 208, 231, 295
troubleshooting 191
a twist on raspberry cheesecake 258–9
recipes 18–19, 43–4, 71–2, 98–9, 162, 187–8, 212–13, 235, 237–8, 258–9, 278–9, 299–300
red-hot poker *see Kniphofia*
redcurrant 77, 156, 182, 186, 207
redwing 20, 259, 281–2, 301
reptiles 74–5
    see also slow worms; snakes
Rhodochiton atrosanguineus 222
rhododendron 105, 152, 180
rhubarb 48
    forced 15–16, 38, 64, 65, 66
    harvesting 38, 65, 93, 126
    planting 13, 38, 275, 277
    rejuvenation 298
    rhubarb and vanilla cake 98–9
    'Timperley Early' 66
    'Victoria' 66
robinia 78
robins 44, 72, 99, 259, 279, 281–2, 300–1
rocket 108, 157
    harvesting 254, 274, 296
    sowing 208, 234, 274
    sweet 174
root flare 263
Rosa viii, 62, 135, 142, 144–5, 153, 163, 171, 175, 190, 214, 216, 249
    bare-root 10, 271
    climbing 62, 178, 192, 295
    deadheading 147, 229
    diseases 192–3, 217, 251
    English 35
    floribunda 35, 62, 148
    hybrid tea 35, 62, 148
    planting 10, 271

pruning 11, 35–6, 180, 201
R. 'Burgundy Ice' 144–5
R. rugosa 6, 247
rambling 180, 201, 206
shrub 11, 36, 262
suckers 78
support 178, 179
rose chafers 190, 214
rose hips 148, 229, 247, 291
rosebay willowherb 48
rosemary 38, 42, 140, 176, 251, 256, 291
rot 298
rowan 279, 280, 281
Rudbeckia 59, 118, 175, 199, 223, 249
    R. fulgida 'Goldsturm' 226
runner bean 92, 125–7, 143, 182–3
    harvesting 208, 233, 254
rust disease 216, 251
rye, grazing 254

sage 42, 104, 115, 123, 185, 256, 297
salad leaves 42, 70, 108, 182
    autumnal 252
    container-grown 96–7, 132
    hardening off 129
    harvesting 93, 126, 130, 183, 185, 254, 296
    oriental 233, 234
    planting 126, 208
    for small spaces 140
    sowing 154, 233, 234
    winter 232, 274, 296
Salix caprea 54
Salvia 151, 172, 177, 241, 249
    S. nemorosa 'Ostfriesland' 172
    S. 'Purple Majesty' 121
sambuca 62
santolina 123
Sarcococca 5, 28, 58, 272, 293
sauce, Malabar spinach 187–8
savory 256
scaevola 150

scent 112, 174
scilla 52
sea holly 7
seaweed feed 47–8, 131, 154, 305
second flushes 180
sedum *see Hylotelephium*
seedbeds 17
seedlings
　hardening off 92
　moving 86
　potting on 85
　pricking out 89, 118
　self-sown 86
　sun protection for 91–2
　thinning out 128
seeds
　collecting/saving 177, 200, 202
　failed 48–9
　germination 48–9
　sorting 292
　sowing 6–7, 58, 88
　watering 57
　see also specific species, sowing
self-seeding plants 144, 192
Senecio cineraria 226
September 219–42
shade 105, 146, 216–17
shallot 38, 40, 42
　'Matador F1' 18
shield bugs 133
shiso 155
shredders 8–9
shrubs
　bare-root 61, 292
　planting 58, 61, 247, 271, 273, 292
　pruning 11, 89, 294
　see also specific shrubs
Siberian bugloss 84
sidalcea 251
Silene vulgaris 137
silt 75
silver ragwort 226
silver wattle 123

siskins 20
Skimmia 248
　S. japonica 'Rubella' 267
slow worms 20, 73–4, 133, 213, 259
slug pellets 101
slugs 89, 117, 140, 251, 253, 257,
　292
smoke bush *see Cotinus*
snails 89, 117, 251, 260–1, **261**, 292
　common chrysalis 261
　common door 261
　garden 261
　great pond 164
　hedge 261
　round 261
snakes, grass 74, 75, 133
snakeshead fritillary 81–2
snow 9, 28, 52, 266, 286
snowdrop 1, 3, 6, 26, 33, 44, 55, 62,
　272
　autumn-flowering 179
　naturalising 227
soil
　acidic 14–15, 29, 56, 75, 166, 256,
　　283
　alkaline 15, 166
　clay 75
　hygiene 283
　improvers 61
　loam 159
　microorganisms 159
　organic matter 75, 81, 267
　sandy 75
　temperature 53, 57
　testing 15
　water retention 81
Solomon's seal 111, 146
Sorbus 'Joseph Rock' 271
soup, red lentil and leek 71–2
spades 8
sparrowhawk 134
sparrows, house 87, 99, 188
spearmint 184

spiders 239–40
  false widow 240
  garden 239–40, 260
  pale crab 240
  wasp 240
  zebra 240
spinach 38, 42
  'Atlanta' 209
  harvesting 183, 185, 254
  'Perpetual' 209
  sowing 93, 125, 209, 233
  'Winter Giant' 209
spring 62, 85, 142, 170
  see also April; March; May
spring onion 92
  Welsh 132–3
spurs 11
squash 158, 181, 185–7
  harvesting 236, 254
  savoury squash rolls 278–9
  storage 298
stag's horn sumach 78
staking 263
starlings 163, 280, 281
Steinernema feltiae 217
stock 167
strawberry 67, 140, 154, 182
  care for 128–9
  container-grown 132
  harvesting 156, 220
  planting 65, 93, 126, 210, 233
  problems 191
  runners 154, 210
Sturgeon, Andy 199
suckers 78
summer 142, 169–93, 195–217
  see also August; July; June
summer solstice 142
sunflower 88
supporting plants 86, 143, 150,
  158–60, 159, 178–9, 263
swallows 100, 134, 142, 188, 189,
  213

swede 167, 254, 276, 276, 282
  'Best of All' 276
  'Brora' 276
  'Ruby' 276
sweet box see Sarcococca
sweet pea 80, 88, 112, 143, 153, 175,
  179, 183, 204, 271
  deadheading 191
  planting out 87, 118–19
  sowing 32, 274
sweetcorn 64, 124, 130, 208, 212
swifts 100, 134, 142, 164, 188, 189,
  213
Swiss cheese plant 224
sycamore 22
Symphyotrichum 6, 59, 117, 118, 174,
  245
  S. 'Little Carlow' 174
  see also Aster
Syringa 78, 111–12, 152
  S. 'Katherine Havemeyer' 112
  S. vulgaris 'Primrose' 111

tabbouleh 212–13
tadpoles 74, 100
Tagetes 92
tarragon, French 155, 209, 278
tayberry 77
tellima 26
tender plants 222, 227
  annuals 155–6, 170
  cuttings 115, 229
  overwintering 241, 245, 251, 266,
    271–2, 274, 283, 286
  planting out 151
  potted 283, 286
  pruning 123
thrush 44
  song 72
thuja 204–5
thunderstorms 170, 196
thyme 42
tiarella 105

tidying up 180
tillering 76
Titchmarsh, Alan 30–1, 41, 61, 95–6, 122, 149, 157, 191, 229, 250, 293
tits 19, 300
    blue 21, 87, 99, 134, 163–4, 241
    great 21, 72, 99, 163
    long-tailed 20
toads 100, 105, 163, 189, 238–9, 259
    common 74
tobacco plant 174
tomato 37–8, 42, 48, 68–72, 92, 155
    AGM varieties 69–70
    beefsteak 68
    blossom end rot 152, 181
    'Brandy Boy' 71
    cherry 68, 96–7
    cherry plum 68
    cordon 96–7, 124, 153, 159–60, 186
    'Favorita' 70
    feeding 187
    'Gardener's Delight' 96–7
    growing tips 69
    halting 252, 274
    hardening off 129
    harvesting 183, 185, 207, 220, 232–3
    'Ildi' 97
    late 232
    'Olivade' 70
    pinching out 153, 186
    planting out 114, 126, 154–6, 255
    plum 68
    pot-grown 96–7, 132
    potting on 65, 125, 127, 129
    ripening aids 181
    'Rosella' 96–7
    salad 68
    sowing 38, 125
    'Suncherry Premium' 70
    'Sungold' 70, 96–7

    training 159–60, **159**, 185–6
    watering 152, 158
top dressing 95
Tophill, Frances 34–5, 205
topiary 150, 251
tortillas, broad bean 162
training plants 185–6
tree fern 272
trees 21–2, 54
    bare-root 61, 275
    planting 61, 247, 263, 271
    staking 263
    see also specific trees
Trillium 83, **83**
    T. erectum 83
troubleshooting 75–8, 165–8, 240–2
    autumn 262–3, 282–3
    spring 103–5, 138–40
    summer 191–3, 216–17
    winter 22–4, 47–50, 303–5
tubers 85, 87, 161–2
Tulip 52, 62, 82, 140
    dwarf 227
    naturalising 227
    planting 228, 248–50, 271
    T. 'Abu Hassan' 250
    T. 'Estella Rijnveld' 250
    T. 'Jan Reus' 82
    T. 'Prinses Irene' 82, 250
    T. 'Queen of Night' 250
    T. 'Spring Green' 250
turnip
    harvesting 14, 254, 296
    problems 167, 282
    sowing 154, 182

Urtica dioica 99, 122–3, 164

vanilla and rhubarb cake 98–9
vegetables *see* growing your own fruit and veg; *specific vegetables*
Verbena 192
    V. bonariensis 199

Veronica 171
  V. 'Marietta' 171
Viburnum 49–50, 63, 291
  V. davidii 217
  V. opulus 271, 281
  V. tinus 29, 152
Vinca major 146
vine weevils 91, 182
vinegar 235
Viola 27–8, 85
  V. 'Sorbet Morpho' 28
Virginia creeper 252, 294
voles 99, 189
  bank 73

wallflower 29, 152, 182, 207, 248,
  282
wasps 100, 133, 189, 192, 260
  parasitic 262
water boatmen 20, 45, 164
water hog-louse 165
water iris 87
water lily 87, 167, 204
watering 88–9, 117, 158, 178
  bedding plants 151
  greenhouses 152
  pots 88–9, 120, **120**, 131, 151, 165,
    171, 178
  seeds 57
waxwings 20, 280–2, **280**
weather 52–3, 80–1, 142–3, 170–1,
  220–1
  autumn 244–5, 266–7
  spring 108–9
  summer 196–7
  winter 2–3, 26–7, 286–7
wedding cake tree 246
weeds 37, 64, 67, 92, 127, 138–9, 154,
  207, 291
  annual 138

dandelions 48, 103
lawn 76
nettle 99, 122–3, 137, 164
summer 182
see also specific weeds
weigela 151–3, 180
whitefly 253
wildflowers 136–7
wildlife 72–5, 163–5, 238–40, 279–82
  autumn 259–61
  spring 99–102, 133–7
  summer 188–91, 213–15
  winter 19–21, 44–7, 300–3, **302**
  see also specific species
willow 62, 77
willow moss 167
Wilson, EH 110
wind 2, 3, 27, 109, 143
wind-rock 262
winged spindle 267
winter 80, 271
  see also December; February;
    January
winter aconite 4
winter-flowering plants 272
wintersweet 290
wisteria 11, 135, 163, 204–6
witch hazel 3, 26, 293
wood anemone 137
woodlice, common 47
wormcasts 22–3, 58
worms 22–3, 101, 267, 300, 305
wraps, mooli 18–19
wreaths 290–1
wren 21, 281, 300

yew 86, 201, 251

zinnia 88, 151, 175, 183, 200